Gene E. Likens
F. Herbert Bormann
Robert S. Pierce
John S. Eaton
Noye M. Johnson

# BIOGEOCHEMISTRY
## of a Forested Ecosystem

Springer-Verlag
New York   Heidelberg   Berlin

Gene E. Likens
Section of Ecology
and Systematics
Cornell University
Ithaca, New York 14853/USA

John S. Eaton
Section of Ecology
and Systematics
Cornell University
Ithaca, New York 14853/USA

F. Herbert Bormann
School of Forestry
and Environmental Studies
Yale University
New Haven, Connecticut 06520/USA

Noye M. Johnson
Department of Earth Sciences
Dartmouth College
Hanover, New Hampshire 03755/USA

Robert S. Pierce
Northeastern Forest Experiment
Station
U.S. Forest Service
Durham, New Hampshire 03824/USA

Photograph on the cover by courtesy of Judith Meyer.

Library of Congress Cataloging in Publication Data

Main entry under title:

Biogeochemistry of a forested ecosystem.

Bibliography: p.
Includes index.
1. Forest ecology—New Hampshire—Hubbard Brook
Valley. 2. Biogeochemistry—New Hampshire—Hubbard
Brook Valley. I. Likens, Gene E., 1935–
QH105.N4B56       574.5'264       76-50113

ISBN 0-387-90225-2 Springer-Verlag New York

ISBN 3-540-90225-2 Springer-Verlag Berlin Heidelberg

# Preface

About 15 years ago we began the Hubbard Brook Ecosystem Study with the development of an ecosystem model and the conception of a method whereby some major parameters of the model could be directly measured under field conditions. The method, called "the small watershed technique," allowed measurement of input and output of chemicals and the construction of ecosystem nutrient budgets. Although the Hubbard Brook study of nutrient cycling originated with ideas developed by F. H. Bormann and G. E. Likens, its early growth was aided by the remaining authors of this volume—Robert Pierce, forest hydrologist; Noye Johnson, geochemist; and John Eaton, forest ecologist. Donald W. Fisher of the United States Geological Survey also cooperated in the early phases of the project and provided numerous data on the chemistry of precipitation and stream water. Particular credit is due the United States Forest Service, whose scientists chose the Hubbard Brook Valley as a hydrologic study site, selected particular watersheds for intensive measurement, carried out a variety of basic hydrologic studies, and in general cooperated with us in many ways to make the Hubbard Brook Ecosystem Study a reality.

The initial part of the ecosystem study was concerned primarily with nutrient flux and cycling and it was done slowly and deliberately. The entire effort during the first few years of study was carried forward by three of us at

Dartmouth College with the cooperation of the United States Forest Service. At that time we had no precedents to follow because similar, comprehensive studies of natural ecosystems had not been done. We reasoned that it would be best to first construct a solid base of studies on nutrient–hydrologic interactions upon which subsequent studies could be built. In this regard, we were fortunate to rather quickly determine quantitative nutrient budgets for replicated ecosystems. These results gave us guiding insight into the function of natural ecosystems and into the development of future lines of research.

Nevertheless, we felt that slower growth would be more manageable and would allow for substantial interaction among all senior investigators to insure proper coordination and development of the overall study. Generally, the method we used to guide the growth of the Hubbard Brook research proposals was as follows: First, based on our own perceptions or feedback from ongoing studies, from cooperating scientists, or from outside advisors, we recognized research problems that were timely and particularly pertinent to our overall research goals. Some of the studies were launched under our direction, others were brought to the attention of an established investigator working in that research area. We then sought a mutually satisfactory arrangement to allow the investigator to work at Hubbard Brook. From the beginning we have called attention to the kinds of information a cooperative study might produce, but always we have encouraged individuality in the design and execution of research. *We deem this individual research freedom one of the greatest assets of the Hubbard Brook Study.* Individuality in selection of problems and conduct of research is also encouraged among graduate students at Hubbard Brook. Not only does this contribute to the intellectual ferment and sound growth of the Hubbard Brook Study, but as educators, we feel this approach absolutely necessary if the Hubbard Brook Study is to contribute to graduate education.

Over the last 15 years the study has grown to include not only hydrology and input–output chemistry but also aspects of the structure, function, and dynamics of the forest ecosystem itself. In addition to studies of undisturbed ecosystems, entire ecosystems have been experimentally manipulated to allow comparative study of undisturbed and perturbed ecosystems. Stream and lake ecosystems within the Hubbard Brook Valley have been studied in detail and the connection between the forest ecosystem and these interlinked aquatic ecosystems is now the focus of considerable attention.

Approximately 50 senior scientists and scores of graduate students have carried out studies at Hubbard Brook during the last 15 years. Over 200 publications have resulted from this work.

We now plan to order some of this information in ways we consider useful both to the scientists concerned with the theory of biogeochemical cycles and the structure, function, and development of forested ecosystems, and to the land-use specialists who are concerned with the production of a variety of benefits, goods, and services from northern hardwood ecosystems.

We have developed two volumes so far and a third is planned for the future. This, the first volume, presents a detailed examination of the biogeochemistry of an undisturbed, aggrading, second-growth northern hardwood forest at Hubbard Brook. Ecosystems similar to this cover a large area of northern New England and New York. Major emphasis in this book is on the physical aspects of nutrient and hydrologic flow through the ecosystem and nutrient budgetary. Not only are we concerned in this book with the presentation of data and conclusions that may be drawn from them, but we also discuss methodology as it influences results and we try to share some of our experience at Hubbard Brook in the hope that other workers can avoid pitfalls where we have not been so fortunate. Moreover, we hope that the data presented in this volume can serve as a basis for decision-making and management schemes, including modeling within the northern hardwood forest landscape.

The second volume has as its primary concern the structure, function, and development through time of the northern hardwood ecosystem and is designed for the advanced student in ecology as well as for the ecosystem specialist. It concentrates on the interrelationships among biogeochemical processes, structure, and species behavior within the ecosystem and how these change with time following perturbation. We plan for the information in these two volumes to interrelate and thus draw support for each other. Biogeochemistry is presented more fully in this first volume, whereas biological considerations are dealt with more fully in the second volume.

By no means do these two volumes summarize the

Hubbard Brook Study. In the future we hope to put together a more comprehensive volume that involves contributions from the many cooperating scientists who have brought their particular specialties to bear on the Hubbard Brook Ecosystem.

There are many persons to acknowledge and thank for contributions to such a multidisciplinary study conducted over so many years. Many data were collected during rainstorms, blizzards, intense cold, darkness, and numerous other types of northern New Hampshire weather. Long, often tedious hours in the laboratory and at the computer center were also required. The ideas, devotion, loyalty, and encouragement and genuine enthusiasm of field technicians, laboratory technicians, students, and various other colleagues made the overall Hubbard Brook Ecosystem Study a reality and is beyond a simple "thank you."

For this little book we owe debts great and small to many people: to those whose ideas and encouragement provided an intellectual thrust to move ahead; to those without whose devotion, skills, and endurance there would have been no progress at all; and to those mostly good-natured colleagues whose shafts of humor and criticisms have helped us all to keep a realistic perspective. For all of these things and more, we thank:

J. Aber, B. Ackerman, M. Alexander, T. Allison, H. Art, P. Art, S. Bennett, R. Bilby, J. Bishee, C. Black, A. Bock, H. Bond, B. Bormann, R. Bormann, D. Botkin, E. Brastrup, J. Brink, S. Brown, A. Brush, L. Burckes, T. Burton, D. Buso, P. Cantino, J. Chamberlain, S. Chisholm, B. Coffin, C. Cogbill, B. Corson, W. Covington, M. Cushman, H. Davis, M. Davis, F. deNoyelles, G. deSylva, A. Dominski, J. Duggin, F. Edington, Y. Falusi, A. Federer, N. Fernando, S. Fiance, A. Field, D. Fisher, S. Fisher, L. Forcier, M. Fox, J. Fryer, G. Furnival, J. Galloway, A. Gambell, D. Gerhart, J. Gilbert, W. Glanz, J. Gosz, C. Grant, M. Gross, R. Hame, G. Hart, B. Haywood, J. Hobbie, R. Holmes, S. Holmgren. J. Hornbeck, S. Howe, J. Janak, P. Johnson, M. Jordan, F. Juang, H. Karsten, M. Keller, W. Kent, L. Kivell, J. Koonce, S. Koppes, D. Korbobo, M. Koterba, B. Kuehn, R. Lavigne, T. Ledig, J. Lehman, R. Leonard, V. Levasseur, E. Loker, B. Mahall, J. Makarewicz, P. Marks, W. Martin, D. Masza, K. McConnochie, O. MacGregor, D. McNaught, A. McPhail, J. Melillo, R. Mendelsohn, R.

Miller, J. Meyer, R. Moeller, M. Mollony, D. Mulcahy, R. Muller, J. Murphy, P. Murtaugh, D. Nelson, J. Nobel, R. Patrick, P. Pelton, B. Peterson, N. (Crane) Poole, G. Potter, R. Reifsnider, W. Reiners, R. Reynolds, L. Riggs, J. Roskoski, D. Ryan, M. Schmale, P. Sargent, E. Sayetta, V. Scarlet, W. Schlesinger, J. Schorske, L. Schwartz, M. Sebring, T. Sherry, T. Siccama, D. Smith, W. Smith, P. Sprague, D. Sprugel, D. Stout, S. Strom, F. Sturges, J. Tinker, P. (Fruend) Tinker, J. Thompson, G. Thornton, D. Tiernan, P. Toyryla, K. Turekian, G. Voigt, J. Wallis, B. Walsh, R. Walter, P. Wicklund, R. Withington, G. Whitney, R. Whittaker, T. Wood, G. Woodwell, J. York, and C. Zoltov.

We all acknowledge continued indulgence from our wives (Kay, Chris, Carol, Diane, and Delores) and families for time spent away from them.

Financial support for the field studies was provided by the National Science Foundation and the United States Forest Service. This work was done through the cooperation of the Northeastern Forest Experiment Station, Forest Service, United States Department of Agriculture, Upper Darby, Pennsylvania.

The manuscript was initially formulated while Gene E. Likens was a John Simon Guggenheim Fellow at the Australia National University. This support is gratefully acknowledged.

January 1977                                    G. E. Likens
                                                F. H. Bormann

# Contents

# 1 Ecosystem Analysis

An ecological system has a richly detailed budget of inputs and outputs of energy and matter. Because of the lack of precise information about these relationships and the internal functions that maintain the ecosystem, it is often difficult to assess the impact of human activities on the biosphere. As a result, land-use planners often cannot take into account or even foresee the full range of consequences a project may have. Without full information, the traditional practice in the management of land resources has been to emphasize strategies that maximize the output of some desirable product or service and give little or no thought to the secondary effects. As a result one sees such ecological confusion as an all-out effort to increase food production while natural food chains become increasingly contaminated with pesticides and runoff and seepage waters carry increasing burdens of pollutants from fertilizers and farm wastes. Forests are cut with inadequate perception of the effects on regional water supplies, wildlife, recreation, and esthetic values, and wetlands are converted to commercial use with little concern over important hydrologic, biologic, esthetic, and commercial values lost in the conversion.

For some time it has been evident that a new conceptual approach to the management of resources is desirable. One approach that has received considerable attention during the last decade is to consider entire ecological systems as single interacting units, instead of limiting studies to only a few components of the system. In an experimental forest in New Hampshire we have been conducting a large-scale investigation aimed at supplying this kind of information that is usually lacking about ecosystems.

A vast number of variables, including biologic structure and diversity, geologic heterogeneity, climate, and season, control the flux of both water and chemicals through ecosystems. Clearly, both living and nonliving components of ecosystems are important in defining and regulating the flow of matter within and between ecosystems. Because chemicals tend to circulate from nonliving components to living organisms and back to the nonliving environment, the pathways have been termed biogeochemical cycles. Ecosystems continually exchange (gain or lose) matter and energy with other ecosystems and the biosphere as a whole and, therefore, biogeochemical cycles are vital to the maintenance of the system. Comparisons of biogeochemical data from natural ecosystems with those that have been manipulated by man or otherwise disturbed then provide important information about the functional efficiency or "health" of an ecosystem.

For more than a decade the biogeochemistry of the Hubbard Brook Experimental Forest, New Hampshire, has been the subject of continuous study (Figure 1). Some 6,000 water samples, including rain, snow, stream water, ground water, throughfall, and stemflow, have been collected and chemically analyzed. These extensive data now provide the means and the perspective to identify, isolate, and quantify some of the complex and interlocking biogeochemical processes of a forested ecosystem. As this study shows, seasonal and year to year variations in flux rates and associated biologic responses often require long-term analysis before reliable generalizations can be drawn about such complex natural systems. This book summarizes our understanding of the biogeochemical flux of water and nutrients through an undisturbed, aggrading northern hardwood forest ecosystem, refers to the time scale at which the various biogeochemical processes operate, and relates these data to other ecosystems throughout the world.

We began studies in 1963 to determine the magnitude of the biogeochemical flux and internal cycling of nutrients in northern hardwood forest ecosystems in the White Mountains of New Hampshire. Our ecosystem unit is a watershed or drainage area, with vertical and horizontal boundaries defined functionally by biologic activity and the drainage of water (Bormann and Likens, 1967). Six

**FIGURE 1.** A model depicting nutrient relationships in a terrestrial ecosystem. Inputs and outputs to the ecosystem are moved by meteorologic, geologic, and biologic vectors (Bormann and Likens, 1967; Likens and Bormann, 1972). Major sites of accumulation and major exchange pathways within the ecosystem are shown. Nutrients that, because they have no prominent gaseous phase, continually cycle within the boundaries of the ecosystem between the available nutrient, organic matter, and primary and secondary mineral components tend to form an intrasystem cycle. Fluxes across the ecosystem's boundaries link individual ecosystems with the remainder of the biosphere.

adjacent, small watersheds within the Hubbard Brook Experimental Forest, hereafter referred to as HBEF, with similar vegetation and geology and subject to the same climate, have been used for these studies (Bormann and Likens, 1967; Likens and Bormann, 1972). This series of watershed ecosystems provided the possibility for replication as well as a design for experimental manipulation of entire natural ecosystems.

A conceptual model of these deciduous forest ecosystems was developed to facilitate the quantitative evaluation of the input–output flux and cycling of water and chemicals (Bormann and Likens, 1967; Likens and Bormann, 1972). The continuous flow of energy, water, nutrients, and other materials across the ecosystem's boundaries are considered to be inputs and outputs, which are transported by meteorologic, geologic, and biologic vectors (Figure 1). Meteorologic inputs and outputs consist of wind-borne particulate matter, dissolved substances in rain and snow, aerosols, and gases (e.g., $CO_2$). Geologic flux includes dissolved and particulate matter transported by surface and subsurface drainage, and the mass movement of colluvial materials. Biologic flux results when chemicals or energy gathered by animals in one ecosystem are deposited in another (e.g., local exchange of fecal matter or mass migrations). These input–output categories are therefore defined as vectors or "vehicles" for transport of nutrients, matter, or energy rather than sources; i.e., a leaf blown into an ecosystem represents meteorologic input rather than biologic input.

Within the ecosystem, the nutrients may be thought of as occurring in any one of four basic compartments: (1) atmosphere, (2) living and dead organic matter, (3) available nutrients, and (4) primary and secondary minerals (soil and rock). The atmospheric compartment includes all elements in the form of gases or aerosols both above and below ground. Available nutrients are ions that are absorbed on or in the humus or clay–humus complex or dissolved in the soil solution. The organic compartment includes all nutrients incorporated in living and dead biomass. The woody tissues of living vegetation are considered a part of the living biomass. The primary and secondary minerals contain nutrients that comprise the inorganic soil and rock portions of the ecosystem.

The biogeochemical cycling of elements involves an exchange between the various compartments within the ecosystem. Available nutrients and gaseous nutrients may be taken up and assimilated by the vegetation and microorganisms; some may be passed on to heterotrophic consumers and then made available again through respiration, biologic decomposition, and/or leaching from living and dead organic matter. Insoluble primary and secondary minerals may be converted to soluble available nutrients through the general process of weathering; soluble nutrients may be redeposited as secondary minerals.

**Small watershed approach**

In humid regions chemical flux and cycling are intimately linked to the hydrologic cycle. Hence one cannot measure the input and output of nutrients without simultaneously measuring the input and output of water. The problem usually is that subsurface flows of water, which can be a significant fraction of the hydrologic cycle, are almost impossible to measure.

About 15 years ago it occurred to us that under certain circumstances the interaction of the nutrient cycle and the hydrologic cycle could be turned to good advantage in the quantitative study of an ecosystem. The requirements are that the ecosystem be a watershed underlain by tight bedrock or some other impermeable base. In that case the only inputs would be meteorologic and biologic; geologic input need not be considered because there is no transfer between adjacent watersheds. In humid areas where surface wind is a minor factor, losses from the system are only geologic and biologic. Given an impermeable base, all the geologic output would inevitably turn up in the streams draining the watershed. If the watershed is part of a larger and fairly homogeneous biotic unit, the biologic output tends to balance the biologic input because most animals move randomly into and out of the watershed, randomly acquiring or discharging nutrients. One therefore need measure only the meteorologic input and the geologic output of nutrients to arrive at the net gain or loss of a given nutrient in the ecosystem. Using this approach, it has been possible to obtain quantitative budgets for most of the major chemical nutrients. The flux of some nutrients with a gaseous phase (e.g., N and S) is more difficult to evaluate, but the ecosystem approach has enabled us to make realistic assumptions and estimates about some of these unmeasured gaseous components. We have used this approach at Hubbard Brook in an intensive study of six contiguous watersheds within the HBEF (Figure 2). These watershed ecosystems are all tributary to Hubbard Brook.

## The Hubbard Brook ecosystem

The HBEF was established in 1955 by the United States Forest Service as the principal research area for the management of watersheds in New England. The name of the area is derived from the major drainage stream in the valley, Hubbard Brook. Hubbard Brook flows generally from west to east for about 13 km (Figure 2) until it joins with the Pemigewasset River, which ultimately forms the Merrimack River and discharges into the Atlantic Ocean. Water from more than 20 tributaries enters Hubbard Brook along its course. Mirror Lake, a small oligotrophic lake, discharges into Hubbard Brook at the lower end of the valley.

### Location

The HBEF is located near West Thornton within the White Mountain National Forest of north central New Hampshire. Coordinates of 43°56′ N, 71°45′ W bisect the area. The Atlantic Ocean is about 116 km to the southeast.

**FIGURE 2.** The Hubbard Brook Valley showing the Hubbard Book Experimental Forest with its gaged watersheds, weirs, weather stations, roads, experimental treatments, and drainage streams tributary to Hubbard Brook.

## Climate

Although the climate varies with altitude, it is classified as humid continental with short, cool summers and long, cold winters (Trewartha, 1954). The climate may be characterized by (1) changeability of the weather, (2) a large range in both daily and annual temperatures, and (3) equable distribution of precipitation. HBEF lies in the heart of the middle latitudes and the majority of the air masses therefore flow from west to east. During the winter months these are northwesterlies and during the summer the air generally flows from the southwest. Therefore, the air affecting HBEF is predominantly continental. However, during the autumn and winter, as the colder polar air moves south, cyclonic disturbances periodically move up the east coast of the United States providing an occasional source of maritime air. The mean air temperature in July is $19°C$ and in January is $-9°C$ (Federer, 1973). A continuous snowpack develops each winter to a depth of about 1.5 m. Occasionally, mild temperatures in midwinter partly or wholly melt the snowpack. A significant microclimatologic feature of this area is that even the uppermost layer of the forest soils usually remains unfrozen during the coldest months because of the thick humus layer and a deep snow cover (Hart et al., 1962).

## Area, topography, and aspect

The HBEF covers an area of 3,076 ha and ranges in altitude from 229 to 1,015 m. The experimental watershed ecosystems (watersheds 1–6, Figure 2) range in size from 12 to 43 ha and in altitude from 500 to 800 m. These headwater watersheds are all steep (average slope of 20–30%) and face south. The experimental watersheds have relatively distinct topographic divides. The height of the land surrounding each watershed ecosystem and the area have been determined from ground surveys and aerial photography.

## Geology

The geologic substrate, outcrops of bedrock and stoney till, in the Hubbard Brook Valley was exposed some 12,000–13,000 years ago when the glacial ice sheet retreated northward (Likens and Davis, 1975). Bedrock is derived from highly metamorphosed sedimentary rocks of

**PHOTOGRAPH 1.** *Aerial view of Hubbard Brook.* Hubbard Brook Experimental Forest, West Thornton, New Hampshire (area within –·–symbol, approximately 3,076 ha). Experimental watersheds shown by dotted boundaries (see Figure 2). Watershed 2 was cleared of all woody vegetation in 1965 and sprayed with herbicides for three successive years to prevent regrowth. No forest products were removed. Watershed 4 was cut in alternating strips in three phases over a 4-yr period beginning in 1970. Picture shows watershed 4 when two-thirds of the strips were cut and harvested. Watershed 101 was cleared and products were removed in one operation in 1970. All other watersheds are untreated as yet and serve as references. Dark areas on the ridges and peaks generally are conifers and the lighter areas are deciduous northern hardwoods.

the Littleton formation and the granitic rocks of the Kinsman formation (Billings, 1956). The bedrock of watersheds 1–6 is the Littleton formation, which in this area is made of highly metamorphosed and deformed mudstones and sandstones. It is medium to coarse grained and consists of quartz, plagioclase, and biotite with lesser amounts of sillimanite. Much of the area of the experimental watersheds is covered with glacial till derived locally from the Littleton formation. The geologic substrate is considered watertight and losses of water by deep seepage are minimal (Likens et al., 1967).

## Soils

Soils are mostly well-drained spodosols (haplorthods) of sandy loam texture, with a thick (3–15 cm) organic layer at the surface. Most precipitation infiltrates into the soil at all times and there is very little overland flow (Pierce, 1967). This is because the soil is very porous, the surface topography is very rough (pit and mound, mostly from windthrown trees), and normally there is little soil frost.

Soil depths are highly variable but average about 0.5 m from surface to bedrock or till. Soil on the ridges may consist of a thin accumulation of organic matter resting directly on the bedrock. In some places, impermeable pan layers at depths of about 0.6 m restrict vertical water movement and root development. The soils are acid (pH ≤ 4.5) and generally infertile.

## Vegetation and fauna

The vegetation of the HBEF is part of the northern hardwood ecosystem, an extensive forest type that extends with variations from Nova Scotia to the western Lake Superior region and southward along the Blue Ridge Mountains (Braun, 1950; Küchler, 1964; Oosting, 1956). Classification of mature forest stands as northern hardwood ecosystems rests on a loosely defined combination of deciduous and coniferous species that may occur as deciduous or mixed deciduous–evergreen stands. Principal deciduous species include, beech *(Fagus grandifolia)*, sugar maple *(Acer saccharum)*, yellow birch *(Betula alleghaniensis)*, white ash *(Fraxinus americana)*, basswood *(Tilia americana)*, red maple *(Acer rubrum)*, red oak *(Quercus borealis)*, white elm *(Ulmus americana)*; the principal coniferous species are hemlock *(Tsuga canaden-*

**PHOTOGRAPH 2.** *A spodosol soil developed on till at the HBEF.* The uppermost dark layer, an organic layer called the forest floor, rests on the mineral soil. Immediately below the forest floor is a bleached and chemically leached horizon, which grades into a zone of accumulated organic matter and iron and aluminum oxides. At still lower depths, the soil varies from partially weathered to unweathered mineral components. The presence of an obvious leached horizon is intermittent within the HBEF.

*sis),* red spruce *(Picea rubens),* and white pine *(Pinus strobus)* (Braun, 1950). At Hubbard Brook, the vegetation is characteristic of a developing, northern hardwood forest ecosystem. The forest was logged in 1910–1919 but has not been disturbed since (Bormann et al., 1970) and there is no

evidence of fire in the history of these stands. The forest is
unevenly aged and well stocked, primarily with sugar
maple *(Acer saccharum)*, American beech *(Fagus grandifolia)*, and yellow birch *(Betula alleghaniensis)*. Red
spruce *(Picea rubens)*, Balsam fir *(Abies balsamea)*, and
hemlock *(Tsuga canadensis)* are prominent on north-fac-
ing slopes, on ridge tops, on rock outcrops, and along the
main channel of Hubbard Brook. The forest has a basal
area of about 24 m²/ha (Bormann et al., 1970). About 100
species of shrubs and herbs may be found in the watershed
ecosystems (Siccama et al., 1970). Net primary productivity and living biomass accumulation for the forest averaged about 1,040 g/m²·yr and 360 g/m²·yr dry weight,
respectively, during 1956–1965 (Whittaker et al., 1974).
Forest floor accumulation averaged about 33 g/m²·yr dry
weight for a 55-yr-old forest (Covington, 1976).

Fauna common to northern hardwood forests occur
within the Hubbard Brook Valley. More than 90 species of
birds have been observed (Likens, 1973) and snowshoe
hare *(Lepus americanus)*, beaver *(Castor canadensis)*,
red fox *(Vulpes fulva)*, black bear *(Ursus americanus)*,
and whitetail deer *(Odocoileus virginianus)* are present.
The White Mountains generally have a low population
density of deer because of severe winters and high hunting
pressure (Siegler, 1968).

**Drainage streams**

Headwater streams draining the watershed ecosystems are
small and perennial. Flow ranges from zero during infrequent summer droughts to hundreds of m³/ha·day during
snowmelt and storm events. Two of the watersheds, 3 and
4 (42 and 36 ha, respectively), are larger than the others
and have maintained flow during every summer of obser-
vation (Figure 2).

From June to October the streams are heavily shaded
by the forest vegetation. As a result daily water temperatures do not vary more than a degree or two Celsius. The
annual temperature range in these streams is from 0 to
about 18°C. The stream water is normally saturated or
supersaturated with dissolved oxygen and the pH is less
than 6.

The stream bed is covered by organic debris, fine sand,
gravel, cobbles, and small boulders. There are occasional
bedrock exposures in the stream channel and these are
preferred sites for gaging weirs. Numerous dams of
organic debris occur throughout the length of the stream;

**PHOTOGRAPH 3.** *Hubbard Brook at an elevation of about 215 m.* The characteristic boulder substrate and "stair-step" nature of this stream are evident.

the stream therefore has a stair-step appearance of alternating "waterfalls" and pools. The stream channel comprises about 1–2% of the area of each watershed ecosystem.

Additional details concerning the topography, climate, geology, and biology of the HBEF are given elsewhere (Bormann et al., 1970; Holmes and Sturges, 1973; Likens et al., 1967; Johnson et al., 1968; Likens, 1973; Marks, 1974; Northeastern Forest Experiment Station, 1972; Siccama et al., 1970; Whittaker et al., 1974). The results of our earlier biogeochemical studies have been reported by Bormann and Likens (1967), Likens et al. (1967), Fisher et al. (1968), Johnson et al. (1968, 1969), Juang and Johnson (1967), Bormann et al. (1969, 1974), Pierce et al. (1970), Likens et al. (1971), Likens and Bormann (1972), Fisher and Likens (1973), Hobbie and Likens (1973), Marks (1974), Sturges et al. (1974), Hornbeck et al. (1970, 1975a,b), Burton and Likens (1975), and Likens and Bormann (1975). In addition to providing basic data on ecosystem function and the ecological relationships between structure and function, these biogeochemical data provide a baseline for judging the effects of human manipulations of forested landscapes (e.g., Likens et al., 1970; Pierce et al., 1972).

The next chapters, Hydrology through Nutrient Cycles, present detailed information about the long-term biogeochemistry of the HBEF. These sections summarize our current understanding of the flux and cycling of chemicals in the watershed ecosystems at Hubbard Brook. An attempt is made to synthesize results and to interpret and comment on outstanding problems, solved and unsolved.

# 2 Hydrology

Because of the vital role of water as a transporting agent, chemical solvent, and catalyst, quantitative data on hydrology are of paramount importance in understanding the biogeochemistry of an ecosystem. The United States Forest Service has monitored and maintained accurate records of precipitation and streamflow for watersheds of the HBEF since 1956.

**The water-year**     Selection of a suitable water-year is a primary consideration for hydrologic analyses. The ideal water-year is that successive 12-month period that most consistently, year after year, gives the highest correlation between precipitation and streamflow. In such watersheds as the HBEF, streamflow is largely dependent on (1) precipitation; (2) the amount of water storage opportunity in the soil and thereby the amount of water stored in the soil at any time; and (3) the amount of water stored in the snowpack.

After a large array of linear regressions for successive 12-month periods of precipitation and streamflow had been examined, the water-year beginning on 1 June and ending on 31 May was chosen. This water-year exhibited correlation coefficients of about 0.99. This high correlation can be explained by several phenomena: (1) Although evapotranspiration each summer leaves the soil in variable states of water content, autumn rains completely replenish water depleted during the summer. (2) Water storage opportunity for the watershed as a whole is on the order of about 10–15 cm; and (3) the addition of between 20 and 30 cm of water to the soil during melting of the snowpack is sufficient to fully recharge the soil. The selection of 1 June water-year is also advantageous because this corresponds to the period when foliage is getting established. This allows for a separation of the water-year into periods essentially coincident with growing and dormant seasons of the vegetation.

**Precipitation**   Because precipitation serves as a major vehicle of nutrient input into the ecosystem, its accurate measurement is of prime importance in an evaluation of any biogeochemical cycle. Precipitation is measured within the area of the experimental watersheds by a fixed network of standard United States Weather Bureau precipitation gages, approximately one for every 13 ha, in conjunction with continuous weighing and recording collectors. The Thiessen polygon method (Thiessen, 1911) is used to determine amounts of areal precipitation over entire watersheds.

On the average 129.5 cm of water falls on each unit area of watershed annually. Of this some 62%, or 80 cm, becomes streamflow and the remainder, about 49 cm, is lost as vapor through evapotranspiration (Table 1). The evapotranspiration is estimated as the difference between precipitation and streamflow for the water-year because the geologic base of the watershed is watertight (see the section on Deep Seepage, below).

There are on the average 111 precipitation days (> trace) per year at Hubbard Brook, or about two per week. Dividing the average annual precipitation of 129.5 cm by 111 days gives an average intensity of 1.2 cm per precipitation day.

**Table 1.**   Average Annual Hydrologic Budget for the Hubbard Brook Experimental Forest

|  | cm/unit area $\pm s_{\bar{x}}$ | Percent of total |
|---|---|---|
| 1956–1974 |  |  |
| Input (precipitation) | 129.5 ± 5.1 | 100 |
| Output (streamflow) | 80.1 ± 5.0 | 61.9 |
| Output (evaporation and transpiration) | 49.4 ± 0.79 | 38.1 |
| 1963–1974 (period of biogeochemical study) |  |  |
| Input (precipitation) | 132.2 ± 6.8 | 100 |
| Output (streamflow) | 83.3 ± 6.6 | 63 |
| Output (evaporation and transpiration) | 48.9 ± 1.0 | 37 |

**PHOTOGRAPH 4 a–d.** *Precipitation collectors.* Various types of precipitation collectors used at the HBEF. Collectors are located in clearings within the forest to avoid contamination. No overhead obstruction is allowed above an angle of 45° from the opening of the collector. Collectors are positioned at least 2 m above the ground. (**a**) Gages for quantitatively monitoring amounts of precipitation (rain and snow). *Left:* standard United States Weather Bureau nonrecording gage with 20.3-cm orifice. *Right:* weight-activated

c

d

recording gage. Alter-type windshields reduce wind velocity around gage opening to provide a sample equivalent to deposition on ground. A mixture of ethylene glycol and methanol are added to gages in winter to convert snow to liquid state. (**b**) A polyethylene apparatus used for collecting rain for chemical analysis consisting of a 28-cm diameter funnel, tubing, 2-liter reservoir, and vapor barrier. Samples are retrieved and a clean apparatus installed at 1-week intervals. (**c**) An automatic sampler for the collection and separation of wet and dry depositions for chemical analysis. A sensor on the lid activates the mechanism to move the lid from one bucket to the other at the beginning of wet or dry periods. (**d**) Plastic garbage can used to collect snow for chemical analysis. Samples are retrieved and a clean collector installed at 1-week intervals.

The concept of a normal or average water-year is of limited utility. Although the long-term average was 129.5 ± 15.8 cm/yr during 18 yr of record (1956–1974), in only 9 yr has precipitation been within ±10 cm of this value (Figure 3). The mean deviation from the average precipitation is 12.2% at Hubbard Brook and is less than the expected mean variability, about 16%, for similar rainfall areas of the world (Conrad, 1941). In the driest water-year on record, 1964–1965, precipitation was 95.1 cm, whereas in the wettest year, 1973–1974, it was 185.7 cm. These values represented the extremes of wet and dry conditions relative to precipitation; they were 27% less and 43% greater, respectively, than the long-term average at Hubbard Brook. This range of 90.6 cm emphasized the need for a long-term record to understand the ecosystem's hydrologic flux. If the biogeochemistry of our ecosystems had been based upon either of these years alone, our estimates of meteorologic input and stream-water output of nutrients would have been substantially misleading. As it turned

**FIGURE 3.** Annual precipitation and streamflow for undisturbed forested watersheds in the Hubbard Brook Experimental Forest. The horizontal lines indicate the average values during 1956–1974. $P$, mean precipitation; $R$, mean streamflow; $ET$, mean evapotranspiration; $ET = P - R$; solid bars show annual streamflow; open bars show annual precipitation. Streamflow from manipulated or disturbed watersheds is not included in these results. Each water-year is from 1 June to 31 May.

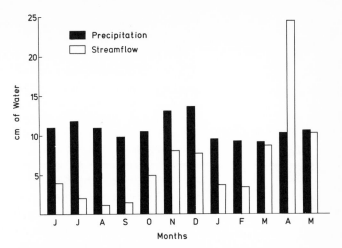

**FIGURE 4.** Average monthly precipitation and streamflow for the Hubbard Brook Experimental Forest during 1956–1974.

out, during the period of our biogeochemical studies, 1963–1974, annual precipitation averaged 132.2 ± 14.9 cm/ yr and this period included both the wettest and driest years of record. Despite the occurrence of both "extreme" wet and dry years during the period of our study (1963–1974), the standard deviation of the mean precipitation was only 11% and the range varied by less than twofold. This relative "stability" of the precipitation for the HBEF imparts stability to the chemical budgets, as well, because the two are interdependent.

In any particular year monthly amounts of precipitation may show random extremes but for the longer term the monthly pattern becomes quite regular (Figures 4 and 5).

**FIGURE 5.** Average monthly precipitation (solid line) and streamflow (broken line) for the Hubbard Brook Experimental Forest during 1963–1974, the period of the biogeochemical study. The vertical bars are ± one standard deviation of the mean.

**Table 2.**  Annual Input of Water as Snow or Rain for the Hubbard Brook
Experimental Forest

| | Percent of annual time period | Percent of annual water input |
|---|---|---|
| Snow | 41 (mid-Nov.–mid-April) | 30 (38.9 cm) |
| Rain | 59 (mid-April–mid-Nov.) | 70 (90.6 cm) |

That is, the average amount of precipitation is relatively constant on a monthly basis throughout the year. On the average, December is the wettest month and March is the driest, but on the average no month deviates widely from the overall monthly average of 10.8 cm.

The seasonal effects of temperature on precipitation (rain or snow) as well as the structure and function of the biomass (e.g., interception of raindrops by the forest canopy and litter) are important in lowering the erosive potential of precipitation. During the winter most of the precipitation falls as snow with a much smaller potential energy of impact. Indeed, some 30% of the annual precipitation falls as snow (Table 2). The proportion of the year when snow is generally expected is larger than this (41%; Table 2) but some rain occurs in the winter months.

**Streamflow and evapotranspiration**

Streamflow is measured continuously throughout the year by automatic recorders at stream-gaging stations, which include either a V-notch weir or a combination of V-notch and San Dimas flume anchored to the bedrock at the base of each watershed. Heating units in the ponding basin above the V-notch permit accurate measurements of streamflow in the coldest weather. In contrast to the monthly uniformity of precipitation, most of the streamflow at the HBEF occurs in spring when the accumulated snowpack melts (Figures 4 and 5). Some 54% of the annual streamflow occurs during March, April, and May, with more than 30% in April alone. During the summer months streamflow is very low, particularly during August and September (Figures 4 and 5). During this period water loss from the ecosystem occurs primarily through evapotranspiration. A secondary, minor peak in streamflow occurs in November and December when transpiration has virtually ceased after the loss of the deciduous leaves, and with somewhat greater autumn rainfall. At that time soil water deficits are eliminated and most of the rain moves through the soil and into the streams. Winter streamflow is diminished as precipitation accumulates as a snowpack.

**PHOTOGRAPH 5.** *Stream gaging weir.* Tandem arrangement of modified San Dimas flume and weir for monitoring streamflow on watershed 6. The metal flume on the left measures high flows; the 90° sharp edged, V-notch weir on the right measures low flows. The gaging station is built on bedrock so that all streamflow is channeled first through the flume and then into the weir. Recording instruments in shelters for both flume and weir monitor streamflow continuously. Propane burners are used during the winter to prevent freezing. Large particulate matter transported by the stream during high flows is caught either in the screen as the flow cascades from the flume over the trough, bypassing the V-notch, or in the ponding basin behind the V-notch. Gaging stations similar to this or utilizing a V-notch weir only have been established on eight experimental watersheds in the HBEF (see Figure 2).

On an annual basis, evapotranspiration remains relatively constant despite major fluctuations in precipitation. This is in contrast to annual streamflow, which is highly correlated with the amount of precipitation (Figure 6). This suggests that on an annual basis evapotranspiration is fairly constant, that it has first call on precipitation, and that once evapotranspiration is satisfied the remainder of the precipitation goes to streamflow. In a general way this appears to be true, but a more detailed analysis suggests a more complex situation. Streamflow during the months of June through September has an exponential relationship to precipitation, starting slowly and rising after about 45 cm (Figure 7). A simulation of evapotranspiration during the growing season, June through September (Figure 7), shows that evapotranspiration gradually increases to about 50 cm of precipitation and then levels off. Although these

**FIGURE 6.** Relationship among precipitation, streamflow, and
evapotranspiration for the Hubbard Brook Experimental Forest
during 1956–1974. The regression lines fitted to these data are $Y = b
+ ax$, where $Y$ = annual streamflow or evapotranspiration in $10^6$l/ha,
$b = Y$ intercept, $a$ = slope, and $x$ = annual precipitation in $10^6$l/ha.
The equations for each line are: streamflow, $Y = 0.96x -4.45$ ($F$ ratio
of 0.0001***; correlation coefficient, 0.99); evapotranspiration, $Y =
0.038x +4.49$ ($F$ ratio of 0.33; correlation coefficient, 0.24).

data are rough, because we do not know precisely the
effect of stored soil water conditions at the end of May and
at the end of September on streamflow and evapotranspiration during June through September, they suggest a
certain reasonable logic about the hydrologic behavior.

The relationship between evapotranspiration and precipitation underscores the powerful regulating role that the
living ecosystem plays in the hydrologic cycle. During the
active growing season, evapotranspiration (mostly transpiration) is an important factor in regulating the amount of
streamflow. Evapotranspiration then not only acts as a
regulator for the hydrologic cycle but also contributes to
tighter nutrient cycles by diminishing streamflow that car-
ries nutrients out of the ecosystem.

Seasonal streamflow varies by orders of magnitude,
whereas annual streamflow varies by no more than twofold and is usually less. Because the chemical flux is so
intimately tied to the hydrologic cycle, the effects of sea-

sonal hydrologic events have corresponding effects on the seasonal nutrient budgets. In contrast, the annual hydrologic budget, with its relatively small variability, imparts long-term stability to the associated annual chemical budgets, as is shown in subsequent sections.

A factor of prime importance in regulating erosion and stream-water chemistry at the HBEF is that the soils have a high infiltration capacity. Because the soil is very porous, virtually all incident precipitation, rain or snowmelt, percolates through the soil even in the largest storms. Surface runoff in the conventional sense is not manifested at HBEF except in a few boggy areas and in channels (Pierce, 1967). The downhill motion of water toward stream beds is by subsurface interflow. The soil usually remains unfrozen during the coldest months because of the thick humus layer and a deep snow cover (Hart et al., 1962). The surface topography is rough (pit

**FIGURE 7.** Relationship among precipitation, streamflow, and evapotranspiration during the months of June through September for watershed 3 of the Hubbard Brook Experimental Forest during 1958–1974. Evapotranspiration (open circles) is determined as the difference between the amount of precipitation and streamflow for 1 June to 30 September. The solid line is fitted by eye. The dashed line represents evapotranspiration as computed from data on precipitation and air temperature using a United States Forest Service hydrologic model, which includes an estimate of the change in total storage of soil water. C. A. Federer, personal communication, (1976).

and mound), which also facilitates infiltration rather than overland flow. Because the geologic substrate is essentially watertight and losses of water by deep seepage are minimal, almost all water has intimate contact with the biologic and inorganic components of the soil as it moves downhill toward the stream channel. Frozen conditions in the stream beds add still further to the control of particulate matter losses in stream water as erodability in winter months normally is less than that during the growing season (Bormann et al., 1974).

**Deep seepage**  One of the key features of the HBEF that enables a close accounting of water budgets is the watertight bedrock. Precipitation entering a watershed must leave by three avenues: (1) evaporation from leaf and soil surfaces, or both, to the atmosphere; (2) deep seepage, including lateral transfer through soil and bedrock across watershed boundaries and vertical movement through deep geologic strata; and (3) streamflow. Direct measurement of the first two is extremely difficult, if not impossible, with present technology and expertise. The latter, however, can be measured with fair degree of accuracy in many watersheds.

If there are strong indications that a given watershed has impermeable bedrock such that deep seepage is minimal and streamflow can be measured adequately, then the other major component of the water budget, evapotranspiration, can be derived quantitatively as the difference between streamflow and precipitation. Reasonable determinations of these components of the water budget for an entire water-year are feasible providing the water-year begins when storage of soil water is consistent each year.

There is substantial evidence that leads us to believe that the experimental watersheds at the HBEF have reasonably watertight bedrock and therefore have negligible water loss through deep seepage. The following is presented in support of this contention.

(I) From a report of a geological survey conducted at the HBEF, we highlight the findings of Bradley and Cushman (1956) regarding bedrock and watertightness:
    (a) Piezometric surface divides tend to coincide with the shape of the topography.

(b) Unconsolidated deposits lie entirely within valleys which are rimmed by walls of bedrock.

(c) The Littleton formation (an intensely metamorphosed originally sedimentary rock) is crossed by several sets of joints but openings are not large. These may extend one to a few meters below the surface but their size and number decrease with depth.

(d) Only relatively small quantities of water occur in and move through these cracks.

(e) They (Bradley and Cushman) concur with Fowler-Billings and Page (1942) and Moke (1946) that there is no evidence of significant faulting in the HBEF.

(f) Solution cavities, such as may occur in limestone or marble, do not exist because the bedrock formations are not soluble.

Based on these findings, Bradley and Cushman conclude that it is improbable that measurable quantities of ground water move from one watershed to another or that measurable quantities of ground water escape from the watershed through major fault zones or solution openings.

(II) Assuming that two small adjacent watersheds have reasonably similar precipitation, vegetative cover, soil, and evapotranspiration, then streamflow on a unit area basis should also be similar. If there is considerable deep seepage, however, then streamflow from the two watersheds is likely to be widely dissimilar. A comparison of streamflow among watersheds 1, 3, 5, and 6 (11.8, 42.4, 21.9, and 13.2 ha, respectively) for 12 water-years was made using an analysis of variance. The forest on watersheds 2 and 4 was cut in the late 1960s and therefore these two watersheds cannot be used in the long-term, simultaneous comparisons of streamflow with watersheds 1, 3, 5, and 6.

In spite of wide variability in yearly streamflow during the period, e.g., 496 mm to 1401 mm, there was no significant difference in streamflow among watersheds 1, 3, 5, and 6. The calculated $F$ value is 0.24, whereas the tabular $F$ value at the 1% level is 4.26. Therefore, the probability that there are significant differences in streamflow between watersheds is extremely small. Because streamflow is virtually the same for each watershed, we conclude that the chances of deep seepage are negligible.

(III) As stated previously, if water loss through deep seepage were negligible, then evapotranspiration should approximate precipitation minus streamflow (runoff) for the common water balance equation: $ET = P - RO \pm \Delta S$, where $P$ is precipitation, $RO$ is streamflow, and $\Delta S$ is the change in storage of soil water. Over the course of an entire year we assume that the change in storage of soil water is zero. Hart (1966) compared $P - RO$ values for watersheds 1 and 3 (HBEF) for 5 yr with calculated values of potential evapotranspiration based on the method by Thornthwaite (1948). Potential evapotranspiration, $PE$, is defined as the evaporation that would occur from a large fully wetted surface; that is, one with an unlimited water supply (Sellers, 1965). The average wateryear values for watersheds 1 and 3 for a 5-yr period (1958–1963) were 56.2 cm for $PE$ and 53.2 cm for $P - RO$. There are several empirical methods for estimating $PE$ (Hamon, 1961; Penman, 1956; Thornthwaite, 1948) and $PE$ calculations by each of these vary slightly; in general however, such values of $PE$ for HBEF are reasonably close to 56 cm. Actual evapotranspiration should be slightly less than $PE$ except in rare cases. For practical purposes the values of $P - RO$ at the HBEF can be considered as close approximations of actual evapotranspiration. The remarkable closeness of the values of $P - RO$ and $PE$ indicate that water loss from deep seepage cannot be a significant factor in the water balance at the HBEF.

(IV) A final bit of evidence that argues for the assumption of watertightness is the chloride budget. As far as we know, chloride is not accumulated in sizeable quantities within these watershed ecosystems in living or dead biomass or in the mineral soil. Moreover, chloride-bearing rocks in the HBEF are virtually nonexistent. Therefore, if losses by deep seepage are negligible, then inputs in precipitation should be observed quantitatively in the stream-water outputs from the ecosystem. As is shown in later sections, on a long-term basis (9 yr) outputs of chloride have not been significantly different from inputs at the 0.05 probability level.

**Representative-
ness of the
hydrology of
Hubbard Brook for
northern New
England**

One of the main concerns about small experimental watersheds is whether the research results obtained can be interpreted and extended to other, larger watersheds. Climate, soil, geologic formation, topography, vegetative cover, and past land use all vary from place to place, yet they play important roles in the behavior of water fluxes through a watershed. In light of the influences of these watershed features on waterflow, the question arises: How representative is the hydrology of the HBEF to other areas of northern New England? Fortunately, there is one watershed characteristic that integrates the influences of most landscape features into a single measurable characteristic—streamflow. Considerable insight into the inner workings of a watershed ecosystem can be obtained by continuously monitoring the flow from that watershed.

In a comprehensive study of streamflow characteristics of watersheds in northern New England, Sopper and Lull (1965, 1970) addressed the question of representativeness of small experimental watersheds. They compared streamflow from eight watersheds in Maine, New Hampshire, Vermont, and Massachusetts with watershed 1 at the HBEF.

Data adapted from Sopper and Lull are presented in Table 3. The eight watersheds were chosen on the basis of being less than 259 km$^2$ (100mi$^2$) in size, being 90% or greater forest covered, having streams that were undisturbed by diversion or regulation, and having good to excellent records of precipitation and streamflow—the latter having simultaneous periods of records from 1 December 1956 to 30 November 1962. Annual and seasonal streamflow at the HBEF was very close to the average of the eight regional watersheds and fell within the maximum and minimum values for all periods.

Direct streamflow (that portion of the annual streamflow that rapidly leaves the watershed during and after storms, as opposed to prolonged base flow) for Hubbard Brook was high, as expected, but was less than direct streamflow from two other watersheds, one of which was much larger in size, 249 km$^2$.

Additional comparisons made by Sopper and Lull (Table 3) indicated close similarities in streamflow characteristics between experimental and regional watersheds, such as (1) number and amount of mean daily discharge above a given volume of flow and (2) flow duration (percent of time flow was equalled or exceeded).

In general, streamflow in small watersheds tends to

**Table 3.** Hydrologic Characteristics for Selected Watersheds in Northern New England [a]

| Stream and location | Area km² | Forest cover % | Precipitation cm | Average annual and seasonal water yield, cm | | | | | | | Direct runoff, % of annual |
|---|---|---|---|---|---|---|---|---|---|---|---|
| | | | | Annual | Spring | Summer | Fall | Winter | Growing season | Dormant season | |
| Austin Stream, ME | 236 | 95 | 120.7 | 72.3 | 36.0 | 9.4 | 14.8 | 12.1 | 28.5 | 43.8 | 16.4 |
| Swift River, ME | 249 | 97 | 126.0 | 72.0 | 37.3 | 9.2 | 15.9 | 9.6 | 33.1 | 38.9 | 22.5 |
| Ashuelot River, NH | 184 | 94 | 113.2 | 58.4 | 32.7 | 3.6 | 10.5 | 11.6 | 15.7 | 42.7 | 11.0 |
| N. Br. Contoocook R., NH | 142 | 95 | 129.8 | 66.8 | 34.0 | 3.9 | 13.8 | 15.1 | 18.7 | 48.1 | 8.5 |
| Otter Brook, NH | 109 | 93 | 114.2 | 57.7 | 31.5 | 3.9 | 10.5 | 11.8 | 15.6 | 42.1 | 8.4 |
| Smith River, NH | 223 | 95 | 119.0 | 56.3 | 32.1 | 4.6 | 9.1 | 10.5 | 16.6 | 39.7 | 11.8 |
| Saxtons River, VT | 186 | 92 | 118.1 | 52.2 | 30.5 | 3.4 | 8.2 | 10.1 | 13.2 | 39.0 | 14.4 |
| N. Br. Hoosic River, MA | 104 | 90 | 155.5 | 74.8 | 38.9 | 6.3 | 13.2 | 16.4 | 20.6 | 54.2 | 23.8 |
| Regional average | 179 | 94 | 124.5 | 63.8 | 34.1 | 5.5 | 12.0 | 12.2 | 20.2 | 43.6 | 14.6 |
| Hubbard Brook #1, NH | 0.12 | 100 | 124.6 | 70.4 | 36.9 | 3.5 | 15.3 | 14.7 | 16.7 | 53.7 | 21.3 |

[a] After Sopper and Lull (1965, 1970).

have sharper peaks, higher storm levels, and shorter periods of sustained flow than it does in large watersheds. Such streamflow characteristics in small watersheds may be accentuated in steep, mountainous terrain with shallow to bedrock soils. Watersheds at the HBEF have these characteristics, but it is remarkable and reassuring that streamflow at the HBEF, e.g., in a watershed of $0.12$ km$^2$, closely approximates flow from watersheds that are more than three orders of magnitude greater in size!

The close relationships between stream-water quantity for streams in the HBEF and larger streams throughout northern New England add considerable support to the contention that HBEF streams are representative in many ways and that research results from the HBEF can be useful in extension and application to other areas for management purposes.

# 3 Chemistry

Weekly samples of precipitation and stream water are obtained from the experimental watersheds for chemical analysis. Rain and snow are sampled with continuously open plastic collectors (Likens et al., 1967). Samples of stream water are collected approximately 10 m upstream from each weir in clean polyethylene bottles. This is necessary because water collected from the ponding basin above the weir would be contaminated by calcium and bicarbonate from the cement in the weir itself. The concentrations of dissolved chemicals characterizing a period of time are reported as weighted averages. These averages are computed by summing the amount of chemical from individual samples of precipitation or stream water during the period and then dividing this value by the total amount of water during the period. Nutrient flux across ecosystem boundaries is determined (1) by multiplying the measured concentration of dissolved chemicals in the accumulated composite sample of precipitation by the amounts of precipitation during the interval and (2) by multiplying the average of measured concentrations of dissolved chemicals from stream water samples taken at the beginning and at the end of the interval by the amount of stream water during the period.

It has been shown that sampling according to a standard time series may seriously underestimate or overestimate a highly variable parameter (e.g., Claridge, 1970); however, most of the dissolved chemicals in stream water at Hubbard Brook fluctuate relatively little in concentration so that serious errors are not produced by weekly or even monthly sampling in some cases (Johnson et al., 1969; Likens et al., 1967).

As standard practice, we do not include data from precipitation samples that contain leaves, bud scales, insects, bird feces, or any other obvious foreign particles. Such contamination represents a serious problem, for a variety of substances may leach from these materials as the water comes in contact with them in the collector. Our intent is to have enough collectors in position to provide at least one "clean" sample per interval. This procedure normally works very well.

Details concerning the routine methods used in collecting samples of precipitation and stream water and analytical procedures have been given by Bormann and Likens (1967), Likens et al. (1967), Fisher et al. (1968), Bormann et al. (1969), and Hobbie and Likens (1973).

**Precipitation chemistry**

Precipitation chemistry has been monitored at the HBEF since 1963. To our knowledge this unbroken series is the longest comprehensive record of precipitation chemistry in the United States. Moreover, samples are collected on a weekly or storm basis instead of on the more commonly used monthly interval. Considering the problems of contamination and biogeochemical transformations that can occur in the reservoir of a bulk precipitation collector, experience has shown that sampling intervals of not longer than a week are highly desirable, if not necessary, to obtain accurate data on precipitation chemistry.

### Annual weighted concentrations

The chemical concentration of precipitation is dominated by $H^+$ ions which constitute 70% of the total cationic strength. The remaining 30% is shared among $NH_4^+$, $Ca^{2+}$, $Na^+$, $Mg^{2+}$, and $K^+$ in that order (Table 4).

Sulfate and hydrogen ion are the most prevalent inorganic ions in bulk precipitation at Hubbard Brook (Table 4). On an equivalent basis sulfate is 2.5 times more important than the next most abundant anion, $NO_3^-$, hydrogen ion is 5.9 times more abundant than the next most abundant cation, $NH_4^+$; in both cases these ions far exceed the total equivalency for all other anions and cations. Essentially, then, the incident precipitation at Hubbard Brook may be characterized as a solution of sulfuric and nitric acids at a pH of about 4.1. On a long-term basis the total

**Table 4.** Weighted Annual Mean Concentration of Dissolved Substances in Bulk Precipitation and Stream Water for Undisturbed Watersheds 1–6 of the Hubbard Brook Experimental Forest

| Substance | Precipitation 1963–1974 | | Stream water 1963–1974 | |
|---|---|---|---|---|
| | mg/l | μEq/l | mg/l | μEq/l |
| $H^+$ | 0.073[a,b] | 72.4 | 0.012[a,c] | 11.9 |
| $NH_4^+$ | 0.22[b] | 12.2 | 0.04[b] | 2.2 |
| $Ca^{2+}$ | 0.16 | 7.98 | 1.65 | 82.3 |
| $Na^+$ | 0.12 | 5.22 | 0.87 | 37.8 |
| $Mg^{2+}$ | 0.04 | 3.29 | 0.38 | 31.3 |
| $K^+$ | 0.07 | 1.79 | 0.23 | 5.9 |
| $Al^{3+}$ | —[d] | — | 0.24[e] | 26.6 |
| $SO_4^{2-}$ | 2.9[b] | 60.3 | 6.3[b] | 131 |
| $NO_3^-$ | 1.47[b] | 23.7 | 2.01[b] | 32.4 |
| $Cl^-$ | 0.47[c] | 13.3 | 0.55[c] | 15.5 |
| $PO_4^{3-}$ | 0.008[f] | 0.253 | 0.0023[g] | 0.1 |
| $HCO_3^-$ | ~0.006[h] | 0.098 | 0.92[i] | 15.1 |
| Dissolved silica | —[d] | — | 4.5[b] | — |
| Dissolved organic carbon | 2.4[j] | — | 1.0[k] | — |
| pH | 4.14 | | 4.92 | |
| Total | 8.64 | (+) 102.9 (−) 97.7 | 18.70 | (+) 198.0 (−) 194.1 |

[a]Calculated from weekly measurements of pH.
[b]1964–1974.
[c]1965–1974.
[d]Not determined, trace quantities.
[e]1964–1970.
[f]1972–1974.
[g]1967–1968, 1972–1974.
[h]Calculated from $H^+$–$HCO_3^-$ equilibrium.
[i]Watershed 4 only, 1965–1970.
[j]1973–1974 (Jordan and Likens, 1975).
[k]Watershed 6 only, 1967–1969.

negative equivalent value is not statistically different ($p < 0.05$) from the total positive value (Tables 4 and 7).

Surprising concentrations of dissolved organic carbon are found in rain and snow at Hubbard Brook (Table 4). Precipitation falling on the Hubbard Brook watersheds contains an average carbon concentration of 2.4 mg/l (Jordan and Likens, 1975). As yet we have not completely characterized this organic matter but only a small fraction is dissociated organic acids (Likens et al., 1976a; Galloway et al., 1976). On the basis of a few storm analyses, volatile carboxylic acids, carbohydrates, and tannin–lignin contribute about ≤30% of the organic matter in precipitation at Hubbard Brook. The actual composition of the dissolved organic carbon in precipitation is probably a complex mixture of many organic species, all of which are present in trace amounts. Dissolved carbon in precipitation represents an additional energy source for the ecosystem and may be of some ecological importance. Studies of these relationships are currently underway at Hubbard Brook.

### Origin of ions in precipitation

Cations and anions in precipitation originate from a variety of sources, including oceanic spray, terrestrial dust, gaseous pollutants, and volcanic emissions. How ionic materials find their way into a precipitation collector is an important enigma. The solubility of atmospheric gases, from either natural and anthropogenic origin, in atmospheric water is obvious but the origin of such nonvolatile constituents as Na, Mg, Ca, and K in precipitation is not so obvious. The dispersion of sea salts, such as NaCl and $MgSO_4$, in atmospheric aerosols is one fairly well-known process (Junge, 1963; MacIntyre, 1974). Nevertheless, only a part of the Ca and K in precipitation can be reasonably attributed to marine sources and the "excess" is customarily assigned to unknown "continental sources" (Cogbill and Likens, 1974; Junge, 1963; Granat, 1972). At Hubbard Brook the long-term ratios for ionic Na : Cl are roughly double those anticipated from transported sea salt (cf. Table 4; Juang and Johnson, 1967; Junge, 1963). In contrast, the ionic Ca : Mg ratio is representative of continental precipitation (Eaton et al., 1973; Ericksson, 1952).

One explanation for the excess cations in precipitation is that soil particles and aerosols in the atmosphere may be chemically altered to produce cations. The ubiquitous hydrogen ions in precipitation may react with the entrained soil particles and release cations into solution (see subsequent sections). Therefore, some of the cations found in solution in a precipitation collector may have been derived from an "artificial weathering" reaction within the continuously open (bulk) collector such as we use at Hubbard Brook. Such "weathering reactions" that occur within the precipitation collection reservoirs would not change the total amount of cations input to the ecosystem but might alter their chemical form or underestimate the input of hydrogen ion in precipitation.

Cations and anions in precipitation are approximately balanced. Small errors ($\sim 0.05$ pH unit) in the determination of pH, which has been used to estimate the concentration of hydrogen ion, are sufficient to explain the discrepancy in the cation–anion balance (Table 4). Considering that these long-term averages include the various sampling and analytical errors over an 8- to 11-yr span, the agreement obtained in equivalents is quite good.

It is important relative to quantitative meteorologic inputs to determine whether the dissolved and particulate materials in a bulk precipitation sample have originated outside the ecosystem's boundaries or whether they have merely been circulating within the boundaries. Because the Hubbard Brook watersheds and the surrounding region (1) are almost entirely forested or have other vegetational ground cover, (2) have a long period of snow cover, and (3) have generally humid conditions during other seasons, we believe that relatively little dust arises from within the ecosystem and that our bulk precipitation collectors measure dissolved and particulate matter that has largely originated outside the ecosystem's boundaries.

Bulk precipitation collectors may be very inefficient in collecting dry deposition, particularly aerosols smaller than 1 $\mu$m. Preliminary estimates, based on some measurements of aerosol concentrations and a reasonable deposition velocity, suggest that the dry deposition of Ca, Mg, Na, K, and Cl is probably a small fraction of bulk precipitation inputs at Hubbard Brook. This is not necessarily the case for nitrogen and sulfur. Small aerosol particles and gases of nitrogen and sulfur may be generated from biogenic activity and from the combustion of fossil fuels. Both gases and small aerosols may be transported long distances in the atmosphere. These chemicals therefore exist in the atmosphere at Hubbard Brook in much higher concentrations than that expected from the local environment, and their dry deposition is underestimated by bulk precipitation collectors. This matter is considered more fully in later sections.

### Elevational effects

There is no significant difference in the content of $Ca^{2+}$, $Mg^{2+}$, $Na^+$, or $K^+$ in precipitation at different elevations [from 545 to 758 m MSL (mean sea level)] within the experimental watersheds (Likens et al., 1967). Furthermore, a study done in 1971–1972 indicated no significant difference between the content of $Ca^{2+}$, $Mg^{2+}$, $Na^+$, $K^+$, $NH_4^+$, or $Cl^-$ in precipitation samples from the experimental watersheds at 610 m MSL and those from the Forest Service Headquarters Station at 252 m MSL. However, based upon an analysis of paired samples from these two locations, the concentrations of $SO_4^{2-}$ and $NO_3^-$ were significantly higher ($p < 1\%$ and $p < 5\%$, respectively) at the lower station. The source of this variation may be related to human activity at the lower elevations but this has not been confirmed.

## Acid precipitation

One of the more interesting and ecologically significant findings from these studies was the high acidity of rain and snow at Hubbard Brook (Likens et al., 1972). The average annual weighted pH from 1964–1965 through 1973–1974 ranged between 4.03 and 4.21 (Figure 8); the lowest value recorded for a storm at Hubbard Brook was pH 3.0 and

**FIGURE 8.** Annual weighted mean concentrations in precipitation for the Hubbard Brook Experimental Forest during 1955–1974. The values for 1955–1956 were extrapolated from isopleth maps given by Junge (1958) and Junge and Werby (1958). Note that the ordinate has been compressed. From Likens and Bormann (1974b).

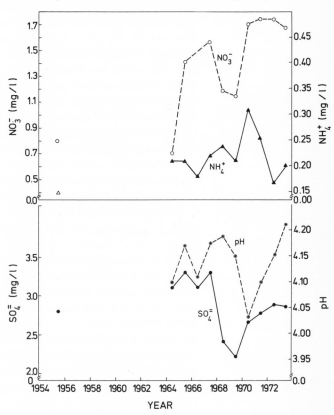

the highest value was 5.95. During the last 5 yr of the study only, no value for weekly precipitation exceeded a pH value of 5.0. Such precipitation is thought to be decidedly abnormal chemically because pure water in equilibrium with atmospheric concentrations of $CO_2$ should have a pH of not less than about 5.6 (Barrett and Brodin, 1955). In other words, the precipitation at Hubbard Brook has a hydrogen ion concentration 50–500 times greater than expected. It is thought that the increased acidity of precipitation is the result of anthropogenic emissions of $SO_2$ and $NO_x$, which are hydrolyzed and oxidized to strong acids ($H_2SO_4, HNO_3$) in the atmosphere (Bolin, 1971; Likens and Bormann, 1974b; Likens et al., 1972). The occurrence of acid precipitation at Hubbard Brook, which is >100 km distant from any large urban–industrial area, emphasizes that one of the major ways man can influence natural ecosystems is through pollution of the atmosphere and that these effects are not limited to areas adjacent to sources of pollution.

Acid precipitation apparently has been falling on much of the eastern United States since at least the early 1950s (Cogbill and Likens, 1974; Likens and Bormann, 1974b). The record for pH of precipitation at Hubbard Brook, although short, is the longest in the United States and provides some information on trends. There has been a slight upward trend in annual weighted average concentrations of hydrogen ion between 1964–1965 and 1970–1971, followed by a downward trend until 1973–1974; overall (1964–1974), however, no trend in concentration is statistically significant (Figure 8). Likewise, concentrations for $SO_4^{2-}$ and $NH_4^+$ vary from year to year but there are no statistically significant trends for the period. In contrast, annual weighted $NO_3^-$ concentrations currently are about 2.3-fold greater than they were in 1955–1956 or in 1964–1965 (Figure 8). An even larger relative increase in nitrate concentration was observed in the central Finger Lakes Region since 1945 (Likens, 1972). These increases in nitrate concentrations are roughly correlated with the increased use of fossil fuels in combustion engines (Likens, 1972) and with increased use of nitrogenous fertilizers, but there may be other explanations. Nevertheless, the increase in nitrate concentration in precipitation in this remote, forested area at Hubbard Brook is indicative of an increasing and widespread distribution of atmospheric pollutants. The ecological implications are difficult to judge.

On the one hand, increased nitrate input into the northern hardwood forest ecosystem and associated aquatic ecosystems may act as nitrogen fertilizer; on the other hand, nitrate contributes to the formation of a potentially harmful increase in acidity of precipitation.

Even though the hydrogen ion concentrations are variable from year to year, the annual input (concentration times volume) in precipitation increased by 1.4-fold during the period 1964–1965 to 1973–1974 (Figure 9). This increased input of hydrogen ion was in sharp contrast to the annual input of all other ions, which remained constant or decreased, except nitrate (Table 5). Based upon a regression analysis, annual nitrate input increased by 2.3-

**FIGURE 9.** Annual hydrogen ion and precipitation input for the Hubbard Brook Experimental Forest. The regression line for hydrogen ion is $Y = 0.033x + 0.819$ where $Y$ is the $H^+$ input in $10^3$Eq/ ha, and $x$ is the year; the significant correlation coefficient is 0.73. Note that the ordinate has been compressed. From Likens et al. (1976b).

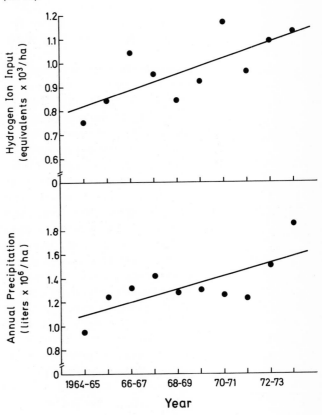

**Table 5.** Dependence of Annual Precipitation Input on Sequential Years for the Hubbard Brook Experimental Forest[a]

| Substance | Slope | Correlation coefficient | Time period |
|-----------|-------|-------------------------|-------------|
| $Ca^{2+}$ | $-0.174$[b] | $-0.83$ | 1963–1974 |
| $Mg^{2+}$ | $-0.046$[c] | $-0.66$ | 1963–1974 |
| $K^+$ | $-0.126$[c] | $-0.66$ | 1963–1974 |
| $Na^+$ | $-0.002$ | $-0.02$ | 1963–1974 |
| $NH^+$ | $0.109$ | $0.57$ | 1964–1974 |
| $H^+$ | $0.033$[c] | $0.73$ | 1964–1974 |
| $SO_4^{2-}$ | $0.838$ | $0.32$ | 1964–1974 |
| $NO_3^-$ | $1.73$[b] | $0.79$ | 1964–1974 |
| $Cl^-$ | $0.300$ | $0.22$ | 1967–1974 |
| Water | $4.95$[c] | $0.72$ | 1963–1974 |

[a]As determined by regression analysis when $Y = ax + b$, where $Y$ = input in kg of dissolved substances/ha or cm of water per unit area, $a$ = slope, $x$ = water-year ($x_1$ = 1963–1964, $x_2$ = 1964–1965, ..., $x_{11}$ = 1973–1974) and $b = Y$ intercept.
[b]Probability of a larger $F$ value $< 0.01$.
[c]Probability of a larger $F$ value $< 0.05$.

fold during the decade; there was no significant increase in annual sulfate input during the period (Likens et al., 1976b).

The increased annual input of hydrogen ion is partially explained by the "increase" in the amount of annual precipitation during the 10-yr period (Table 5); that is, the early years of the decade (1963–1965) were drought years and the last year (1973–1974) was extremely wet (Figure 3). However, close examination of data for individual years shows that factors other than increased amount of precipitation are important (Figure 9). Consequently, annual concentrations alone, even though they are weighted for volume of precipitation, do not accurately reflect trends in the annual input of hydrogen ion.

Sulfuric acid is the dominant acid in precipitation at Hubbard Brook. Based on a stoichiometric formation process in which a sea salt anionic component is subtracted from the total anions (Cogbill and Likens, 1974), sulfate has contributed more than 65% of the acidity in precipitation throughout the decade. However neither the concentration nor the input of sulfate changed significantly during the period of our study (Figure 8 and Table 5). Therefore it is not surprising that the increase in input of hydrogen ion from 1964 to 1974 is not significantly related to the change in input of sulfate. In contrast, the increase in input of hydrogen ion was highly correlated with the increased input of nitrate (Figure 10). The 1 : 1 relationship between annual inputs of hydrogen ion and nitrate is a powerful argument that nitric acid is the crucial variable for explaining the increased input of hydrogen ion during the past 10 yr (Likens et al., 1976b). Although sulfuric acid dominates the precipitation at Hubbard Brook and accounts for most of the hydrogen ion in precipitation, therefore, the increase in input of hydrogen ion during the past decade apparently has been caused by an increase in the annual nitric acid content of precipitation falling on this forested ecosystem.

Precipitation chemistry has changed both qualitatively and quantitatively at Hubbard Brook during the past decade. Absolute concentrations have varied (Figure 8) and relative proportions of the component chemicals have changed (Figure 11). Based upon stoichiometry, the sulfate contribution to acidity dropped from 83 to 66% and nitrate increased from 15 to 30% from 1964–1965 to 1973–

**FIGURE 10.** Relationship between the annual hydrogen ion input and the annual nitrate input during the period 1964–1965 to 1973–1974. The regression line, $Y = 1.07x + 0.631$, where $Y$ is annual input of hydrogen ion in $10^3 Eq/ha \cdot yr$ and $x$ is annual nitrate input in $10^3 Eq/ha \cdot yr$, is highly significant (correlation coefficient of 0.84, and probability of a larger $F$ value is $<0.01$). From Likens et al. (1976b).

**FIGURE 11.** Variation in ionic composition of precipitation for the Hubbard Brook Experimental Forest during 1964–1965 to 1973–1974. From Likens et al. (1976b).

1974. Annual inputs reflect these changes in complex ways. In a further attempt to resolve the relative importance of the various factors that controlled the annual hydrogen ion inputs, a stepwise multiple regression analysis was done to relate the annual hydrogen ion input to a variety of independent variables. An analysis of five independent variables indicated that 86% of the variability in annual hydrogen ion input during the decade was related to annual nitrate input. Six percent of the variability was caused by annual sulfate input, 5% by the input of the sum of all cations minus hydrogen ion, 2% by year, and <0.01% by the annual amount of precipitation.

These data illustrate how economic decisions and activities far removed from this region may influence the dynamics of the northern hardwood ecosystem. Apparently several decades ago, changes in atmospheric chemistry resulting from changes in fossil fuel consumption and disposal of gaseous effluents resulted in decreased pH of precipitation over the northeastern United States. Initially, changes in $H^+$ concentration were related to changes in sulfur chemistry (Likens et al., 1972) but during the last decade, although sulfuric acid dominated the precipitation at Hubbard Brook, increased input of $H^+$ was caused by an increase in atmospheric $NO_3^-$.

The biogeochemical effects of excess acidity in precipitation and its increased input during the past decade at Hubbard Brook are currently under study. Because precipitation is weakly buffered, the presence of small amounts of strong mineral acids, such as $H_2SO_4$, can greatly depress the pH value. Even though only relatively small amounts of acid are involved the potential effect of increased amounts of hydrogen ions on biologic and chemical systems can be large.

Fish kills of major proportions and diminished reproduction of salmonoid fish have been observed and attributed to acid precipitation in areas of similar geology in the Adirondack Mountains of New York (Schofield, 1976), in Canada (Beamish, 1976), and in Scandinavia (Wright et al., 1976). No fish are found in the small headwater streams at Hubbard Brook, but the effect of the acidic precipitation on this distribution is not known. Eastern brook trout occur in the streams at lower elevations in the HBEF.

The first portions of meltwater ($\sim 20\%$) from the snowpack may be significantly more acid than subsequent portions (Hornbeck et al., 1976; Johannessen et al., 1976). This phenomenon may cause a sharp drop in the pH of drainage streams and lakes during the spring and has been blamed for massive fish kills in Scandinavia (Wright et al., 1976). This seasonal change in pH also has been observed in streams and in Mirror Lake within the Hubbard Brook Valley, but it is a very temporal condition of variable intensity. During the winter and with intermittent melting, the snowpack becomes less acid than the incident precipitation (Hornbeck et al., 1976). If the snow melts rapidly there may be a flush of highly acidic water into the stream channels; however, if the snowpack melts more slowly the reduction of pH in stream water is much less pronounced.

Despite seasonal differences in acidity of precipitation, effects of snow accumulation and melt, and variations in residence time of soil water, the acidity of stream water at Hubbard Brook at the level of the gaging weirs is remarkably constant throughout the year (Figure 12). Even though the average monthly concentration for hydrogen ion in precipitation for watershed 6 ranged from 46 to 102 $\mu$Eq/l, average monthly concentrations in stream water from watershed 6 only varied from 10 to 17 $\mu$Eq/l during 1964–1974. The uniformity of stream-water acidity indicates a strong buffering action of the terrestrial ecosystem (Figure 13). The terrestrial ecosystem is therefore very effective in

**FIGURE 12.** Weighted monthly hydrogen ion concentration and pH for precipitation and streamflow for watershed 6 of the Hubbard Brook Experimental Forest during 1965–1974. The vertical lines for each month represent one standard deviation. From Hornbeck et al. (1976).

**FIGURE 13.** General relationships for two seasonal periods at Hubbard Brook. The values are hydrogen ion concentrations in $\mu$Eq/liter. Modified from Hornbeck et al. (1976).

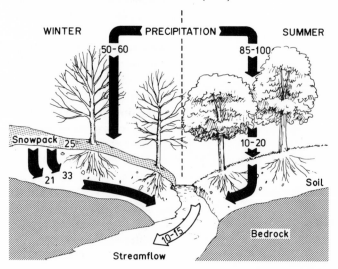

ameliorating the short-term impact of the acid precipitation on the associated aquatic ecosystems.

Our model of the forest ecosystem (Figure 1) suggests a number of functions internal to the forested ecosystem that may be affected by acid precipitation. For example, leaching of substances from the canopy can be speeded up and, in fact, that seems to be the case. Eaton et al. (1973) found that 90% of the $H^+$ ions striking the summer canopy of the forest were consumed within the canopy, presumably releasing an equivalent amount of basic cations. Laboratory studies of seedlings of major tree species at Hubbard Brook (Wood and Bormann, 1974, 1975) indicated increased leaching of cations as the $H^+$ content of artificially applied mist increased.

The podzol soils at Hubbard Brook are already quite acid and the geochemical effects of acid precipitation appear to be minimal (Johnson et al., 1972). However, over the long term, acid precipitation could have an important effect on a variety of soil processes, e.g., nitrogen fixation or nitrification. Theoretically 1 m of rain at pH 4 could leach 50 kg/ha of $CaCO_2$ from the system in drainage water.

Whittaker et al. (1974) and Cogbill (1976) reported a decline in forest growth at Hubbard Brook during the last two decades; in general, however, significant effects of acid rain on forest growth are extremely hard to detect. The forest ecosystem is regulated by a large number of such environmental factors as the amount and distribution of rainfall, temperature, the length of growing season, unusual meteorologic events, the availability of nutrients, and insect and disease outbreaks. It is extremely difficult to separate the effects of these variables from those of acid precipitation. It is entirely possible, however, that acid rain may be reducing forest growth by small and persistent amounts over wide areas or upsetting the ecological balance so as to reduce the capacity of certain species to resist insect or disease attacks. Cumulative effects over a number of years may result in serious deterioration of the forest ecosystem, particularly as acid precipitation is but one of an array of pollutants affecting the landscape (Bormann, 1974).

### Calculation of mass input of chemicals

The meteorologic input of chemicals for the HBEF is calculated by multiplying the measured concentrations of each ion in the weekly precipitation sample (mg/l or mEq/l) by the volume of precipitation incident over the watersheds for that week (l/ha). The meteorologic input of ions is then expressed in units of mass (or chemical equivalency) per unit area per week. The weekly inputs may then be summed conveniently into monthly, seasonal, or yearly increments.

### Meteorologic input—dissolved substances

The average annual input of dissolved inorganic substances in rain and snow was calculated to be 72.5 kg/ha or 1,320 Eq/ha during 1963–1974 (see Table 10). The input of specific elements is dealt with in a later section when nutrient budgets are discussed. During 1964–1965, one of the drought years, the total input in equivalents was exceptionally low (23% less than the long-term average); during the wet year, 1973–1974, the ionic input was 24% greater than the long-term average (Figure 14). Overall the flux of chemicals in precipitation to these forested ecosystems is substantial, both as nutrients and as weathering

**FIGURE 14.** Annual input of dissolved inorganic substances in precipitation to the experimental watersheds 1–6 of Hubbard Brook Experimental Forest. TDIS = total dissolved inorganic substances.

agents. This is particularly true for $H^+$, $NH_4^+$, $NO_3^-$, $Cl^-$, $PO_4^{3-}$, and $SO_4^{2-}$ and to a lesser extent for $K^+$, $Ca^{2+}$, $Mg^{2+}$, and $Na^+$.

### Throughfall and stemflow

The water that reaches the forest floor is of vastly different chemical content than the incident precipitation. Only about 87% of the incident precipitation actually reaches the ground of the forest ecosystem during the growing season. The remainder is evaporated (or absorbed) directly from the surface of leaves, branches, and stems (Leonard, 1961). As the precipitation passes through the forest canopy its chemistry is altered significantly (Table 6). Throughfall and stemflow (about 5% runs down the stems of trees) during the growing season are greatly enriched in potassium (about 91 times), phosphorus (18 times), magnesium (15 times), and calcium (10 times). All other dissolved substances also increase, with the exception of hydrogen ion, which is held within the canopy by cation exchange reactions.

**Table 6.** Weighted Average Chemical Composition of Incident Precipitation and Throughfall under Different Size Classes of Sugar Maple, Yellow Birch, and American Beech Trees during 1 June 1969 to 28 October 1969[a]

| Material | Precipitation above canopy | Precipitation under canopy (throughfall and stemflow) | Net change |
|---|---|---|---|
| Ca | 0.16 | 1.59 | +1.43 |
| Mg | 0.03 | 0.45 | +0.42 |
| K | 0.07 | 6.37 | +6.30 |
| Na | 0.06 | 0.14 | +0.08 |
| $NO_3$-N | 0.22 | 0.67 | +0.45 |
| $NH_4$-N | 0.21 | 1.21 | +1.00 |
| Total N | 0.44 | 2.44 | +2.00 |
| $PO_4$-P | 0.0026[b] | 0.15 | +0.147 |
| $SO_4$-S | 0.9 | 5.4 | +4.5 |
| Cl | 0.45 | 1.46 | +1.01 |
| H | 0.087 | 0.010 | −0.077 |
| Organic C | 2.4[c] | 12 | +10 |

[a]Values are in mg/l (modified from Eaton et al., 1973).

[b]Weighted average for 1972–1974.

[c]Weighted average for 1973–1974.

Based on an average carbon concentration of 2.4 mg/l and an average precipitation of $129.5 \times 10^5$ l/ha·yr, some 31 kg/ha·yr of dissolved organic carbon are added to the Hubbard Brook ecosystems each year in direct precipitation. As the precipitation passes through the forest canopy it is enriched in organic carbon (Table 6). Throughfall and stemflow during the growing season contain some 52 kg carbon per hectare (Eaton et al., 1973), so an additional 21 kg carbon per hectare must be leached from the forest canopy and added to that already present in the precipitation. These relationships probably are much more complicated than this, because the microflora of the leaf and bark surfaces of the vegetation may metabolize the more labile, dissolved organic substances in precipitation and in turn secrete other organic substances.

There are various sources for the dissolved substances in throughfall and stemflow, but their relative importance is not entirely clear (Eaton et al., 1973): (1) Some nutrients are contained in the incident precipitation; (2) some of the nutrients may have been impacted aerosols that are washed off by the incident precipitation. These two portions should be considered as a part of the chemical flux from outside the ecosystem. Similarly, (3) nutrients with a normal gaseous phase, which were incorporated directly on or into the plant (for example, fixation of C by photosynthesis, absorption, and reaction with $SO_2$ or $NH_3$) and then removed from tissues by incident precipitation should be considered as meteorologic inputs. The remaining nutrients in throughfall and stemflow, which have been leached from the vegetational tissues or associated microflora, are nutrients cycling primarily within the intrasystem cycle of the ecosystem (Figure 1) and should not be considered part of the meteorologic flux. We have not been able to quantify the amounts of chemicals transported along each of these pathways for the ecosystems at Hubbard Brook. Some specific data are presented later in the discussion of nutrient cycles at Hubbard Brook.

**Stream-water chemistry**
The chemistry of stream water has been monitored in watersheds 1–6 of the HBEF since 1963. Experiments that involved cutting of the forest vegetation were undertaken in watershed 2, starting in November of 1965, and in watershed 4, starting in October of 1970. After these dates, therefore, data on stream-water chemistry from these two watersheds were not included in the results presented here.

## Dissolved substances

The Hubbard Brook Forest ecosystem brings about both qualitative and quantitative chemical changes in rain and snowmelt water as it passes through the system (Table 4). Water enters the system as a dilute solution of sulfuric and nitric acid (pH $\sim$ 4) but leaves the system containing primarily neutral sulfates and nitrates (pH $\sim$ 5). Calcium and sulfate dominate the stream-water chemistry, but sodium, magnesium, and aluminum also are relatively important cations on an equivalent basis. Sulfate is four times more abundant than the next most abundant anion, $NO_3^-$, in stream water. The balance between cations and anions in stream water is very good over the 11-yr span (Tables 4 and 7). Annually the smaller value (on an equivalent basis) was always within 10% of the larger value.

The ionic strength of stream water (0.20 mEq/l) is twice that of the incoming water (0.10 mEq/l). A principal factor here is the concentration effect (distillation) of evapotranspiration. As water is lost from the system by evapotranspiration, chemicals in solution tend to be concentrated into a smaller volume of water. Based upon an average annual evapotranspiration loss of 38% (Table 1), the concentration factor would be 1.6, which would fall short of the

**Table 7.** Arithmetic Mean of Annual Weighted Concentrations of Dissolved Substances in Bulk Precipitation and Stream Water for Undisturbed Watersheds 1–6 of the Hubbard Brook Experimental Forest during 1963–1974

| | Precipitation, mg/l | | | | Stream water, mg/l | | | |
|---|---|---|---|---|---|---|---|---|
| Substance | $\bar{x}$ | $s_{\bar{x}}$ | $n^a$ | $s_{\bar{x}}/\bar{x}$, % | $\bar{x}$ | $s_{\bar{x}}$ | $n^a$ | $s_{\bar{x}}/\bar{x}$, % |
| $Ca^{2+}$ | 0.17 | 0.02 | 11 | 11.8 | 1.65 | 0.06 | 11 | 3.6 |
| $Mg^{2+}$ | 0.05 | 0.01 | 11 | 20.0 | 0.38 | 0.01 | 11 | 2.6 |
| $K^+$ | 0.07 | 0.02 | 11 | 28.6 | 0.23 | 0.01 | 11 | 4.3 |
| $Na^+$ | 0.12 | 0.01 | 11 | 8.3 | 0.88 | 0.02 | 11 | 2.3 |
| $Al^{3+}$ | — | — | — | — | 0.23 | 0.01 | 6 | 4.3 |
| $NH_4^+$ | 0.22 | 0.01 | 10 | 4.5 | 0.04 | 0.01 | 10 | 25.0 |
| $H^+$ | 0.0739 | 0.003 | 10 | 4.4 | 0.0126 | 0.0018 | 9 | 14.0 |
| $SO_4^{2-}$ | 2.87 | 0.11 | 10 | 3.8 | 6.23 | 0.11 | 10 | 1.8 |
| $NO_3^-$ | 1.43 | 0.11 | 10 | 7.7 | 1.93 | 0.31 | 10 | 16.1 |
| $Cl^-$ | 0.51 | 0.09 | 7 | 17.6 | 0.54 | 0.02 | 7 | 3.7 |
| $PO_4^{3-}$ | 0.008 | — | 2 | — | 0.002 | — | 2 | — |
| $HCO_3^-$ | — | — | — | — | 1.62 | 0.12 | 5 | 13.5 |
| Dissolved silica | — | — | — | — | 4.43 | 0.11 | 10 | 2.5 |

$^a$years of measurement.

observed concentration factor for chemicals (2.0). However, the concentration factor is not so straightforward as
it appears, for not only do concentrations in stream water
change but the proportions of ionic or dissolved species
change as well, indicating important internal chemical and
biologic reactions. In any case, an additional quantity of
dissolved salts is acquired by the water from within the
ecosystem's boundaries. This process is related to chemical weathering reactions (Johnson et al., 1968) and is
discussed separately.

Appreciable amounts of dissolved organic matter are
found in the undisturbed tributaries to Hubbard Brook.
Fisher (1970) reported an average dissolved organic matter
concentration of 2.34 mg/l during 1968–1969 in Bear
Brook. This would be 1.05 mg of carbon per liter if the
dissolved organic matter were 45% carbon. Hobbie and
Likens (1973) found an average annual weighted concentration of 1.0 mg dissolved organic carbon per liter in
stream water from watershed 6. Most of the stream water
carbon values ranged between 0.3 and 2.0 mg/l, but a
maximum carbon value of 4.8 mg/l was observed during an
extremely high streamflow period (29 July 1969, $34 \times 10^4$ l/
ha·day).

In general the measured stream-water concentrations of
most dissolved substances vary within a narrow range
(less than a factor of two), even though discharge of water
may fluctuate over four orders of magnitude during an
annual cycle (Fisher and Likens, 1973; Johnson et al.,
1969; Likens et al., 1967). This is particularly true for
magnesium, sulfate, chloride, and calcium concentrations.
Sodium and dissolved silica concentrations may be diluted
up to threefold during periods of high streamflow, whereas
aluminum, hydrogen ion, dissolved organic carbon,
nitrate, and potassium concentrations are increased with
increased discharge. Biotic activity within the ecosystem
plays an important role in determining these relationships
for potassium and nitrate in stream water. Nitrate and
potassium are quite sensitive indicators of biologic activity; therefore, stream-water concentrations for these two
nutrients are markedly reduced during periods of plant
growth and increased during periods of vegetation
dormancy.

To empirically describe these variations in stream-
water chemistry we have developed a model that predicts
stream-water concentration in relation to stream discharge
(Johnson et al., 1969). The model is based on some rather

simple and intelligible assumptions. First of all, we assume that two discrete water types exist in our watershed ecosystem; one is represented by relatively dilute water recently added to the system in rain or snowmelt, and the other is older, more salty ground water. Second, we assume that stream water at a given point at any time is some mixture of these two water types. Last, we assume that ground water is added to the stream at a constant rate, whereas rainwater or snowmelt water are added on a variable, day to day basis. Essentially, then, stream water becomes a variable mixture between pure rainwater or snowmelt water and pure ground water, the proportions of which vary from day to day and season to season. During flood periods, say during the spring snowmelt, stream water is composed of nearly pure snowmelt water. During drought periods, say in midsummer, stream water is composed of nearly pure ground water. In spite of these tightly prescribed and perhaps simplistic assumptions, the model does in fact work and serves to reliably and accurately describe quantitative fluctuations in stream-water chemistry on a day to day or season to season basis (Johnson et al., 1969). Within an undisturbed watershed the model predicts quite precisely the changes in stream-water chemistry elicited by a given storm or a given dry period. The

**PHOTOGRAPH 6.** *Winter conditions at the HBEF.* Note that the snowpack has mostly buried the instrument shelter for the gaging weir on watershed 6 (see page 21).

simple premises of the model testify to the remarkable chemical stability and chemical controls inherent in these undisturbed forested ecosystems. It is significant in this regard that the model completely breaks down when the watershed is deforested (Likens et al., 1970).

The maintenance of relatively constant stream-water chemistry is in part a result of the year-long high infiltration capacity and permeability of the forest soils at the HBEF, and in part of a high rate constant for the chemical reactions manifested in the soil zone. Apparently, chemical equilibria between aqueous and solid phases are achieved rapidly through various geologic and biologic reactions in the soil. As a consequence, stream-water chemistry is essentially established in the soil zone or the interflow drainage passages. This is an excellent example of the way in which both the form and rate of chemical flux are modified and tightly regulated by an undisturbed forest ecosystem. Similar constancy of stream-water chemistry has been observed for the major cations in the Coweeta Experimental Forest of North Carolina (Johnson and Swank, 1973).

These examples suggest that the relative independence of stream-water chemistry from stream-water discharge is a rather general feature of undisturbed eastern deciduous forest ecosystems on granitic substrates. The importance of this should not be underestimated, for it demonstrates that the chemistry of regional headwater streams is the product of the natural ecosystem, the control of which may be altered by disturbance. For example, in disturbed forested areas and agricultural lands concentrations of chemicals in stream water usually increase and vary markedly with discharge (e.g., Likens and Bormann, 1974a; Likens et al., 1970; Powell, 1964, pp. 18–19).

Although the concentration of most ions in stream water varies little from year to year (note the standard error of annual arithmetic means, Table 7), the annual weighted concentration of nitrate has been generally increasing during the period 1964–1974 (Figures 15 and 17). For example, the average weighted concentration during 1964–1967 was 0.92 mg/l; during 1967–1969 it was 1.40 mg/l, and in 1969–1972 it was 3.14 mg/l. An explanation for this change in the nitrate concentration is primarily circumstantial and there are several possibilities. The snow cover formed very late in the winter of 1969–1970 and for the first time on record widespread freezing of the soils was observed. Annual stream-water concentration of

**FIGURE 15.** Weighted annual concentration of nitrate in stream water for undisturbed watersheds at Hubbard Brook.

nitrate rose markedly that year (Figure 15). Then, in 1973–1974, widespread freezing of the forest soils occurred again, and again nitrate concentration increased. Various workers have reported that the freezing and thawing of soils promotes nitrification and the mobilization of nitrate to drainage waters (Arefýeva and Kolesnikof, 1964; references cited in McGarity and Rajaratnam, 1973). Therefore, freezing and thawing of the Hubbard Brook soils may have induced nitrification and the subsequent loss of relatively large amounts of nitrate in stream water. However, we have no explanation for the very slow rate of recovery (residual effect) observed during the 2 yr after the freezing occurred in 1969–1970. It might have been expected that if soil frost enhanced nitrification, the nitrate would have been flushed rapidly from the system and in subsequent years nitrate concentrations in stream water would have returned to "normal."

An inverse relationship between nitrate and sulfate concentration in stream water has been observed in undisturbed ecosystems (Figure 16; Likens et al., 1970). This

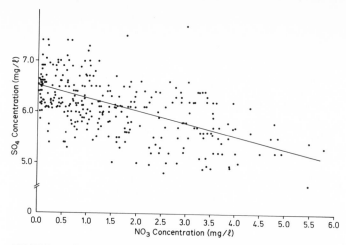

**FIGURE 16.** Relationship between nitrate and sulfate concentrations in stream water. Data were obtained weekly during October through May from 1964 to 1974. The $F$-ratio for this regression line is significant ($p < 0.001$ for a larger $F$-value) and the correlation coefficient is 0.54.

inverse relationship has a strong seasonal feature. Streamwater concentrations of $NO_3^-$ generally reach a maximum during the late winter, whereas concentrations of $SO_4^{2-}$ at that time are at a minimum. Although there has been no consistent decline in stream-water sulfate concentrations since 1968, sulfate concentrations in 1969–1970 were about 10% less than the long-term average and the input of sulfate during 1968–1972 was generally lower than in other years (see Table 10). It may be that part of the increased nitrate concentration in stream water after 1968 is related in some manner to the sulfur biogeochemistry of the ecosystem. This deserves further study.

Another possible explanation for the increasing concentration of nitrate in stream water during the period of our study is that nitrate concentrations in precipitation have also been increasing (Figures 8 and 17). However, there is very poor correlation between the pattern for annual values of nitrate in precipitation (Figure 8) and in stream water (Figure 15); that is, nitrate concentrations in precipitation are relatively low in 1969–1970, whereas this is the peak year for stream-water concentrations.

**FIGURE 17a and b.** Monthly concentrations of nitrate in precipitation (a) and stream water (b) for watershed 6 during the period 1964–1974. The regression line is significant ($p < 0.001$ for a larger $F$-value) and the correlation coefficient is 0.33. ▶

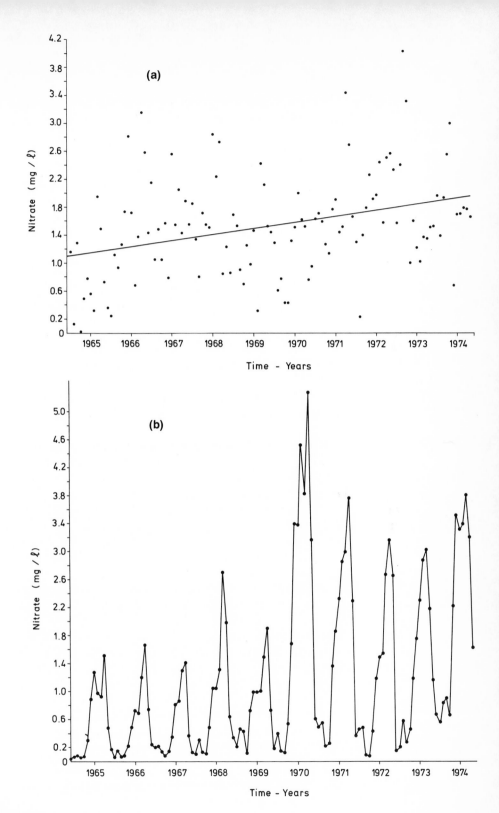

A final complicating factor is that some watersheds show a larger response for nitrate concentrations in stream water than others. For example, the change in nitrate concentration in stream water from 1972–1973 to 1973–1974 was greatest in watershed 1 (40% increase) and somewhat less in watershed 6 (28% increase) and watersheds 3 and 5 (25% increase). The most obvious difference between these watersheds is in area: watershed 1, 11.8 ha; watershed 6, 13.2 ha; watershed 5, 21.9 ha; watershed 3, 42.4 ha; even so, a causal relationship is not clear, particularly for the larger watersheds, 3 and 5.

*Calculation of mass output of dissolved substances.* The output of dissolved substances from our watershed ecosystems is calculated in a manner analogous to calculations of input in bulk precipitation as described above. Stream-

**FIGURE 18.** Relationship between annual streamflow and gross output of calcium, sodium, magnesium, and potassium (kg/ha) during 1963–1974 for undisturbed watersheds 1–6 of the Hubbard Brook Experimental Forest. The regression lines fitted to these data are: $Y = b + ax$, where $Y$ = gross output in kg/ha, $b$ = $Y$ intercept, $a$ = slope, and $x$ = annual streamflow in $10^6$ l/ha. The equations for each line are: $Ca^{2+}$, $Y = 0.37 + 1.61 x$ ($F$-ratio of 56.4***; correlation coefficient, 0.93); $Na^+$, $Y = 1.92 + 0.64 x$ ($F$-ratio of 39.8***; correlation coefficient, 0.90); $Mg^{2+}$, $Y = 0.57 + 0.31 x$ ($F$-ratio of 82.2***; correlation coefficient 0.95); $K^+$, $Y = 0.24 + 0.196 x$ ($F$-ratio of 52.0***; correlation coefficient, 0.92).

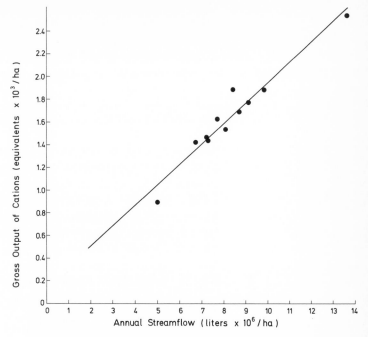

**FIGURE 19.** Relationship between annual streamflow and gross output of cations ($10^3$Eq/ha) during 1963–1974 in undisturbed watersheds 1–6 of the Hubbard Brook Experimental Forest. The regression line fitted to these data is $Y = b + ax$, where $Y$ = gross output of cations in $10^3$Eq/ha, $b$ = $Y$ intercept, $a$ = slope, and $x$ = annual streamflow in $10^6$l/ha. The equation for the line is $Y = 0.15 + 0.18\ x$ ($F$-ratio of 142***; correlation coefficient, 0.97).

water outputs are expressed in mass units (kg) or ionic units (Eq) per hectare of watershed ecosystem.

*Relationship of mass nutrient output to annual streamflow.* Since the chemical concentrations are relatively constant in stream water, the gross annual export of most ions is directly related to the annual streamflow and is highly predictable. The regressions relating gross output of calcium, magnesium, potassium, and sodium to annual discharge are all positive and highly significant (probability for a larger $F$-value $< 0.001$; Figure 18). The total equivalent export for the sum of all cations also can be predicted accurately from annual streamflow data (Figure 19). This is to be expected because the outputs of the major cations in stream water are significantly related to annual streamflow (Figure 18).

The high predictability of mass output in stream water represents one of the major findings of the Hubbard Brook Ecosystem Study. Because of the regulating effect of the forest ecosystem on chemical concentrations in stream water as it emerges from the system, mass output can be predicted simply from a knowledge of annual hydrologic output. Not only is this relationship important in characterizing conditions in an undisturbed forest and thus providing a baseline vital to measurement of any changes resulting from forest manipulation, but it also provides information useful to land-use and hydrologic planners.

*Annual variation in mass output of dissolved substances.* On a mass basis (kg/ha) sulfate and dissolved silica are the dominant substances exported in stream water, whereas on an ionic basis sulfate and calcium dominate. The annual gross output of total dissolved substances (excluding dissolved organic carbon) during 1963–1974 averaged 147.2 kg/ha; excluding dissolved silica, it was 109.6 kg/ha (Table 10). The predominance of sulfate is striking. On the average, the annual gross output of sulfate is 52.8 kg/ha or $1.1 \times 10^3$ Eq/ha, which represents, on the average, 36% of the total dissolved substances or 68% of the anionic equivalency. The annual gross output of total inorganic dissolved substances varies relatively little from year to year (Figure 20). Annual mass export varied by threefold and equivalency by 2.9-fold, which could be largely explained by the range (2.7-fold) in annual streamflow (Figure 3). The largest output of ions occurred during the 1973–1974 water-year (60.4% greater than the long-term average) and the drought year of 1964–1965 had the smallest output (45.4% less than the long-term average; Figure 20). The average annual output of cations during the period 1963–1964 to 1973–1974 was calculated to be $1.65 \times 10^3$ Eq/ha, and the average annual output of anions during 1963–1974 was $1.61 \times 10^3$ Eq/ha. The poorest agreement in equivalents, 11% (difference based on the smaller value), between cations and anions (Figure 20) occurred in 1969–1970 and the best agreement, 1%, was in 1964–1965 and 1971–1972.

Sizeable amounts of dissolved organic carbon are lost from the undisturbed Hubbard Brook watersheds each year in stream water. Based on an average carbon concentration of 1 mg/l (Hobbie and Likens, 1973), about 8 kg/ha of dissolved organic carbon are exported each year.

## Particulate matter

*Measurement of mass output.* In addition to dissolved chemical substances, both organic and inorganic particulate matter may be exported from ecosystems in stream

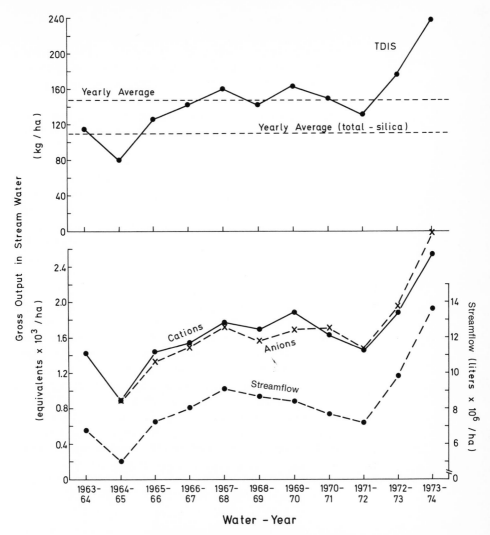

**FIGURE 20.** Annual gross output of dissolved inorganic substances in stream water from undisturbed watersheds 1–6 of the Hubbard Brook Experimental Forest. TDIS = total dissolved inorganic substances.

water. Particulate matter is removed from the watershed as suspended load carried by turbulent water and as bed-load rolled, slid, or bounced along the stream bed. Heavier particulate matter collects in the ponding or stilling basin behind the weir where, over an 8-yr period, it has been periodically dug out, weighted, proportionally sampled, oven dried, and analyzed for dry weight and organic and inorganic content (Bormann et al., 1969, 1974). Suspended particulate matter that passed over the weir was periodically sampled using a 1-mm mesh net and by passing a sample of water through the net and then filtering it through a $0.45\,\mu$m Millipore filter at 40 psi. Samples were then analyzed for dry weight and organic and inorganic content and expressed as concentrations per unit volume of water. These concentrations were used in combination with the hydrologic discharge record to determine the export of materials over the weir. These three components, basin, netted, and filtered, were combined to give total particulate export from the ecosystem (Figure 21).

It is noteworthy that this estimate of total export includes both suspended particulate matter and bedload, whereas most reports of particulate export from ecosystems are concerned with suspended load alone. If the heavier inorganic material that collects behind the weir is considered as a measure of bedload, then bedload constitutes about 55% of the total export.

*Annual loss.* Erosion and transport of particulate matter from these undisturbed, forested watersheds is relatively low even though the watersheds are on steep slopes (12–13°) and are subject to large amounts of precipitation (130 cm/yr). Although the glacial till in these watersheds is relatively resistant to erosion (Hunt, 1967) it is primarily the living and dead biomass that minimizes the loss of particulate matter by regulating the amount, timing, and effect of moving water within the undisturbed ecosystem. Organic debris dams in the stream channels of these head-

**FIGURE 21.** Sampling procedure employed in the measurement of total nutrient losses from a watershed ecosystem. $A$ = water sample for dissolved substances, $B$ = suspended and bed load dropped in basin, $C$ = water sample passed through net for millipore filtration, $D$ = net sample. Total losses = dissolved substances ($A$) + particulate matter ($B + C + D$). After Bormann et al. (1969).

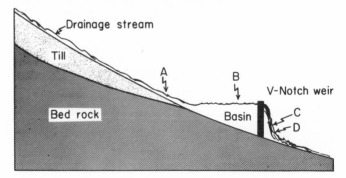

water ecosystems apparently play a major role in minimizing the export of particulate matter.

The undisturbed ecosystems tightly regulate erosion and transport and lose only about 33 $\pm$ 13.4 kg/ha of particulate matter with annual runoff (Table 8). This mate-

**Table 8.** Annual Particulate Matter Output in Kilograms of Oven-Dry Weight Organic and Inorganic Materials per Hectare for Watershed 6[a]

|           | Source of output | Organic | Inorganic | Total |
|-----------|------------------|---------|-----------|-------|
| 1965–1966 | Ponding basin    | 2.12    | 1.77      | 3.89  |
|           | Net              | 0.34    | 0.01      | 0.35  |
|           | Filter           | 1.37    | 1.28      | 2.65  |
|           | Total            | 3.83    | 3.06      | 6.89  |
| 1966–1967 | Ponding basin    | 13.41   | 17.07     | 30.48 |
|           | Net              | 0.39    | 0.01      | 0.40  |
|           | Filter           | 2.72    | 2.95      | 5.67  |
|           | Total            | 16.52   | 20.03     | 36.55 |
| 1967–1968 | Ponding basin    | 3.83    | 5.93      | 9.76  |
|           | Net              | 0.43    | 0.01      | 0.44  |
|           | Filter           | 2.61    | 2.82      | 5.43  |
|           | Total            | 6.87    | 8.76      | 15.63 |
| 1968–1969 | Ponding basin    | 4.61    | 8.31      | 12.92 |
|           | Net              | 0.42    | 0.01      | 0.43  |
|           | Filter           | 2.57    | 2.81      | 5.38  |
|           | Total            | 7.60    | 11.13     | 18.73 |
| 1969–1970 | Ponding basin    | 11.28   | 30.67     | 41.90 |
|           | Net              | 0.40    | 0.01      | 0.41  |
|           | Filter           | 3.30    | 3.69      | 6.99  |
|           | Total            | 14.98   | 34.37     | 49.30 |
| 1970–1971 | Ponding basin    | 2.70    | 2.75      | 5.45  |
|           | Net              | 0.39    | 0.01      | 0.40  |
|           | Filter           | 1.77    | 1.78      | 3.55  |
|           | Total            | 4.86    | 4.54      | 9.40  |
| 1971–1972 | Ponding basin    | 3.66    | 2.52      | 6.18  |
|           | Net              | 0.37    | 0.01      | 0.38  |
|           | Filter           | 1.76    | 1.80      | 3.56  |
|           | Total            | 5.79    | 4.33      | 10.12 |
| 1972–1973 | Ponding basin    | 17.88[b] | 77.40    | 95.28 |
|           | Net              | 0.60    | 0.01      | 0.61  |
|           | Filter           | 10.09   | 13.61     | 23.70 |
|           | Total            | 28.57   | 91.02     | 119.59 |
| 8-yr average | Ponding basin | 7.44    | 18.30     | 25.74 |
|           | Net              | 0.42    | 0.01      | 0.43  |
|           | Filter           | 3.27    | 3.84      | 7.11  |
|           | TOTAL            | 11.13   | 22.15     | 33.28 |

[a]Ponding basin output = particulate matter collected in a ponding basin of the weir upstream from the V-notch; Net output = suspended particulate, >1mm, that passes over the V-notch of the weir; Filter output = suspended particulate matter, >0.45$\mu$m <1mm, that passes over the V-notch of the weir.

[b]Underestimate.

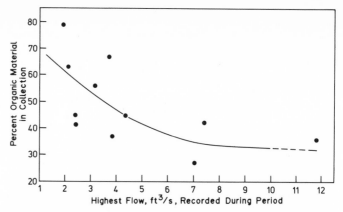

**FIGURE 22.** Percent organic material in basin collections from watershed 6 plotted against the highest flow recorded during the period. One ft³/sec equals 28.3 l/sec. After Bormann et al. (1974).

rial averages about 33 ± 4% (standard error) organic, but the organic : inorganic proportion is strongly influenced by flow rate of the stream, with higher inorganic proportions associated with higher flow rates (Figure 22; Bormann et al., 1974).

*Seasonal variation in erodability.* Although the maximum output of particulate matter occurs during high-discharge periods, summer flows are more effective in exporting particulate matter than are flows of about the same velocity when vegetation is dormant (Bormann et al., 1974). The 30% greater erodability during the summer is probably related to increased biologic activity and decomposition within the ecosystem. In winter the stream channels are often stabilized by ice and snow and, in addition, precipitation generally falls as snow, with low potential energy of impact.

**Particulate matter versus dissolved substance export**

There is a sharp contrast in the response of dissolved substance and particulate matter export to discharge rate of the stream. Concentrations of dissolved substances are relatively little affected by flow rates, whereas particulate matter concentrations are directly and exponentially related to stream discharge (Figure 23). As a consequence, the bulk of particulate matter is moved during storms. Output of dissolved substances is therefore closely related to annual output of water, whereas removal of particulate matter is more of a stochastic process, strongly related to the occurrence of random storms. This is reflected in the fact that during the period 1965–1973 particulate matter output ranged from 7 to 120 kg/ha·yr and was highly correlated with the occurrence of individual storms in a particular year. An analysis of more than 4 yr of data indicated that 86% of the total particulate matter was exported in 1.6% of the total time in the period and with 23% of the water; 16% was exported in 0.0025% of the time period with 0.2% of the water (Bormann et al., 1974).

These relationships have important implications for the export of certain elements. Most of the chemical elements with a sedimentary cycle are exported from the watershed ecosystem primarily in the dissolved form, as shown above; however, the majority of the iron and phosphorus is lost in the form of particulate matter (Table 9). In contrast to other elements, therefore, the output of iron and phosphorus is more directly related to stream discharge rate and is less predictable from annual streamflow. Moreover, phosphorus in particulate form is unavailable to aquatic microorganisms and plants within the stream ecosystem unless it first undergoes decomposition or mineralization. This linkage and the relationships between the losses of particulate and dissolved phosphorus in stream water are currently under study in the HBEF.

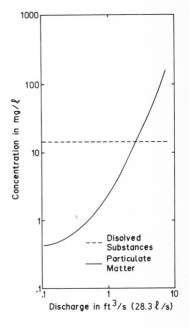

FIGURE 23. General relationships between the concentration of dissolved substances and particulate matter and stream discharge in the Hubbard Brook Experimental Forest. After Bormann et al. (1969).

**Table 9.** Average Annual Gross Output of Some Elements as Dissolved (*D*) and Particulate (*P*) Substances from Watershed 6 of the Hubbard Brook Experimental Forest[a]

| Element | Particulate + dissolved element total, kg/ha | Particulate | | Dissolved | |
|---|---|---|---|---|---|
| | | kg/ha | % of P + D element total | kg/ha | % of P + D element total |
| Aluminum | 3.37 | 1.38 | 40.9 | 1.99 | 59.1 |
| Calcium | 13.93 | 0.21 | 1.7 | 13.7 | 98.3 |
| Chloride | 4.58 | — | 0 | 4.58 | 100 |
| Iron | 0.64 | 0.64 | 100 | — | 0 |
| Magnesium | 3.34 | 0.19 | 5.7 | 3.15 | 94.3 |
| Nitrogen | 4.01 | 0.11 | 2.7 | 3.90 | 97.3 |
| Phosphorus | 0.019 | 0.012 | 63.2 | 0.007 | 36.8 |
| Potassium | 2.40 | 0.52 | 21.7 | 1.88 | 78.3 |
| Silicon | 23.8 | 6.19 | 26.0 | 17.6[b] | 74.0 |
| Sodium | 7.48 | 0.25 | 3.3 | 7.23 | 96.7 |
| Sulfur | 17.63 | 0.03 | 0.2 | 17.6 | 99.8 |
| Carbon[c] | 12.3 | 3.98[d] | 32.4 | 8.35[e] | 67.5 |

[a]Particulate matter losses during 1966–1967 to 1969–1970 are modified from Bormann et al. (1974). Dissolved substance losses are averages during the period 1963–1974.

[b]Assuming that dissolved silica is in the form of $SiO_2$.

[c]Organic only.

[d]Assuming organic matter is 40% C.

[e]Assuming an average carbon concentration of 1 mg/l (Hobbie and Likens, 1973).

Our data show that in the undisturbed forest ecosystem at Hubbard Brook, solution losses are the major force in the geologic process of fluvial denudation. In terms of gross export of mass, dissolved substances—about 165 kg/ha, including dissolved organic matter—are about five times greater than particulate matter losses (Tables 8 and 10).

### The role of debris avalanches in landscape denudation

Our data suggest that headwater watersheds, with their well-developed forests, are gradually lowered in place by the action of solution coupled with slow mass movements. Both of these actions deliver material to the stream, whence it is removed by erosion and transportation. The possibility that infrequent debris avalanches play an equal or greater role in moving material downslope in headwater watersheds should also be considered. For example, Hack and Goodlett (1960) have shown in the Appalachian Mountains of Virginia and West Virginia that infrequent but catastrophic landslides play a major role in removing material from low-order watersheds and that this activity is independent of any restraining influence by the biota. Flaccus (1958a,b) has shown essentially the same thing for the White Mountains of New Hampshire. Our data on stream losses allow a rough comparison of these very different denudational processes.

Flaccus (1958a,b) mapped 543 debris avalanches on aerial photos of the White Mountains. His study area, most of which is heavily forested, is adjacent to the Hubbard Brook valley and contains extensive stretches of the Littleton formation which underlies our watershed. The mapped avalanches occur in about equal proportions on slopes of the Littleton formation and slopes underlain by plutonic rocks. Flaccus examined in detail a range of avalanches, thought to be a fair sample of the total, and calculated the average weight of material moved downslope as 21,800 metric tons. In those slides fresh enough to permit location of the topmost elevation it was found that 100% occurred above the 610 m elevation contour. Within his study we have calculated an area of about 128,000 ha above the 610 m elevation. Using Flaccus' data for numbers and dates of avalanches and average amounts of material moved in each event, we have estimated that 462 kg/ha·yr are moved downslope by debris avalanches occurring above 610 m in the White Mountains (Bormann et al., 1969). This estimate is about triple our estimate of 200 kg/ha·yr for dissolved plus particulate export from headwater watersheds of the HBEF. This suggests that debris avalanches are at least equivalent to the combined action of solution and slower mass movements in lowering the surface of low-order watersheds in the White Mountains.

However, two additional factors must be considered. (1) Disturbances that destroy the biotic stability of the ecosystem, such as fire, can lead to greatly increased losses by solution, mass wasting, and erosion; and (2) the incidence of avalanches per unit area increases sharply with increasing elevation and increasing slope (Flaccus, 1958a,b). In the White Mountains, only 18% of the recorded avalanches occur wholly below 910 m.

At Hubbard Brook it seems unlikely that debris avalanches have occurred within recent times in the area currently occupied by northern hardwood forest. With minor exceptions, elevations within the Hubbard Brook valley are below 910 m and except for a few sites of very restricted area our slope does not meet Flaccus' minimum slope requirement of 25°, nor is there any evidence of recent or old debris avalanches. It seems safe to conclude that debris avalanches have not played a significant role in the denudation of the relatively gentle slopes of the Hubbard Brook watershed, at least during the last millenium.

Apparently, denudation in the Hubbard Brook area is primarily a result of the less dramatic action of solution and erosion in combination with slower mass movements, such as creep, which may deliver materials to the streambeds. This, in turn, suggests that the weathering rind in our watersheds is fairly old. Locally, this is shown by well-developed soil profiles; however, in some places the surface is subject to shallow stirring by burrowing animals and from trees uprooted by the wind.

# 4 Input–Output Budgets

Because our ecosystems are watersheds, the flux of chemicals into the system may be simplified to that in precipitation (meteorologic vector) and the output to that in drainage waters (geologic vector), at least for those nutrients without a prominent gaseous phase (Bormann and Likens, 1967; Likens and Bormann, 1972). Theoretically, then, the difference between annual input ($I$) and output ($O$) for a given chemical constituent tells whether that constituent is being accumulated within the ecosystem, $I > O$; is being lost from the system, $I < O$ or is quantitatively passing through the system, $I = O$. By this definition weathering products do not represent an input to the ecosystem; instead, weathering is considered as a process internal to the ecosystem and a part of the intrasystem cycle (Figure 1; Bormann and Likens, 1967; Likens and Bormann, 1972).

**Annual variation**    We now have, depending on the chemical constituent, from 2 to 11 yr of annual budgets for precipitation input and stream-water output. We are therefore able to estimate with some accuracy mean annual budgets for most of the major ions (Tables 10 and 11). Over the long term there has been considerable variation; in general, however, a net loss of calcium, magnesium, potassium, sodium, sulfate, aluminum, and dissolved silica and a net gain of ammonium, hydrogen ion, and phosphate have occurred from

**Table 10.** Annual Budgets of Bulk Precipitation Inputs and Stream-Water Outputs of Dissolved Substances for Undisturbed Watersheds of the Hubbard Brook Experimental Forest

| Substance (kg/ha) | 1963–1964 | 1964–1965 | 1965–1966 | 1966–1967 | 1967–1968 | 1968–1969 | 1969–1970 | 1970–1971 | 1971–1972 | 1972–1973 | 1973–1974 | Total (1963–1974), kg/ha | Annual mean, kg/ha |
|---|---|---|---|---|---|---|---|---|---|---|---|---|---|
| **Calcium** | | | | | | | | | | | | | |
| Input | 3.0 | 2.8 | 2.7 | 2.7 | 2.8 | 1.6 | 2.3 | 1.5 | 1.2 | 1.2 | 2.0 | 23.8 | 2.2 |
| Output | 12.8 | 6.3 | 11.5 | 12.3 | 14.2 | 13.8 | 16.7 | 13.9 | 12.4 | 15.6 | 21.7 | 151.2 | 13.7 |
| Net | -9.8 | -3.5 | -8.8 | -9.6 | -11.4 | -12.2 | -14.4 | -12.4 | -11.2 | -14.4 | -19.7 | -127.4 | -11.5 |
| **Magnesium** | | | | | | | | | | | | | |
| Input | 0.7 | 1.1 | 0.7 | 0.5 | 0.7 | 0.3 | 0.5 | 0.5 | 0.4 | 0.5 | 0.4 | 6.3 | 0.6 |
| Output | 2.5 | 1.8 | 2.9 | 3.1 | 3.7 | 3.3 | 3.5 | 3.1 | 2.8 | 3.3 | 4.6 | 34.6 | 3.1 |
| Net | -1.8 | -0.7 | -2.2 | -2.6 | -3.0 | -3.0 | -3.0 | -2.6 | -2.4 | -2.8 | -4.2 | -28.3 | -2.5 |
| **Potassium** | | | | | | | | | | | | | |
| Input | 2.5 | 1.8 | 0.6 | 0.6 | 0.7 | 0.6 | 0.8 | 0.6 | 0.3 | 0.5 | 0.8 | 9.8 | 0.9 |
| Output | 1.8 | 1.1 | 1.4 | 1.7 | 2.2 | 2.2 | 2.1 | 1.7 | 1.6 | 2.1 | 2.8 | 20.7 | 1.9 |
| Net | 0.7 | 0.7 | -0.8 | -1.1 | -1.5 | -1.6 | -1.3 | -1.1 | -1.3 | -1.6 | -2.0 | -10.9 | -1.0 |
| **Sodium** | | | | | | | | | | | | | |
| Input | 1.0 | 2.1 | 2.0 | 1.3 | 1.7 | 1.1 | 1.9 | 1.4 | 1.5 | 2.0 | 1.4 | 17.4 | 1.6 |
| Output | 5.9 | 4.5 | 6.9 | 7.3 | 9.1 | 7.6 | 8.0 | 6.5 | 5.8 | 8.1 | 9.8 | 79.5 | 7.2 |
| Net | -4.9 | -2.4 | -4.9 | -6.0 | -7.4 | -6.5 | -6.1 | -5.1 | -4.3 | -6.1 | -8.4 | -62.1 | -5.6 |
| **Aluminum** | | | | | | | | | | | | | |
| Input | a | a | a | a | a | a | a | a | a | a | a | a | a |
| Output | 1.6[b] | 1.2 | 1.7 | 1.9 | 2.1 | 2.2 | 2.2 | 1.8[b] | 1.7[b] | 2.3[b] | 3.2[b] | 21.9 | 2.0 |
| Net | -1.6 | -1.2 | -1.7 | -1.9 | -2.1 | -2.2 | -2.2 | -1.8 | -1.7 | -2.3 | -3.2 | -21.9 | -2.0 |
| **Ammonium** | | | | | | | | | | | | | |
| Input | 2.6[b] | 2.1 | 2.6 | 2.4 | 3.2 | 3.1 | 2.7 | 3.9 | 2.8 | 2.5 | 3.7 | 31.6 | 2.9 |
| Output | 0.27[b] | 0.27 | 0.92 | 0.45 | 0.24 | 0.16 | 0.51 | 0.23 | 0.05 | 0.18 | 0.42 | 3.7 | 0.34 |
| Net | 2.3 | 1.8 | 1.7 | 2.0 | 3.0 | 2.9 | 2.2 | 3.7 | 2.8 | 2.3 | 3.3 | 27.9 | 2.6 |
| **Hydrogen** | | | | | | | | | | | | | |
| Input | 0.85[b] | 0.76 | 0.85 | 1.05 | 0.96 | 0.85 | 0.93 | 1.18 | 0.97 | 1.08 | 1.14 | 10.62 | 0.96 |
| Output | 0.08[b] | 0.06[b] | 0.05 | 0.07 | 0.06 | 0.09 | 0.09 | 0.14 | 0.13 | 0.16 | 0.20 | 1.13 | 0.10 |
| Net | 0.77 | 0.70 | 0.80 | 0.98 | 0.90 | 0.76 | 0.84 | 1.04 | 0.84 | 0.92 | 0.94 | 9.49 | 0.86 |

| | | | | | | | | | | | | | |
|---|---|---|---|---|---|---|---|---|---|---|---|---|---|
| **Sulfate** | | | | | | | | | | | | | |
| Input | 33.7[b] | 30.0 | 41.6 | 42.0 | 46.7 | 31.2 | 29.3 | 34.6 | 33.0 | 43.4 | 52.8 | 418.3 | 38.0 |
| Output | 42.7[b] | 30.8 | 47.8 | 52.5 | 58.5 | 53.3 | 48.1 | 51.1 | 46.8 | 64.0 | 84.7 | 580.3 | 52.8 |
| Net | −9.0 | −0.8 | −6.2 | −10.5 | −11.8 | −22.1 | −18.8 | −16.5 | −13.8 | −20.6 | −31.9 | −162.0 | −14.8 |
| **Nitrate** | | | | | | | | | | | | | |
| Input | 12.8[c] | 6.7 | 17.4 | 19.9 | 22.3 | 15.3 | 14.9 | 21.6 | 21.4 | 26.3 | 30.9 | 209.5 | 19.0 |
| Output | 6.7[c] | 5.6 | 6.5 | 6.6 | 12.7 | 12.2 | 29.6 | 24.9 | 18.7 | 19.2 | 34.8 | 177.5 | 16.1 |
| Net | 6.1 | 1.1 | 10.9 | 13.3 | 9.6 | 3.1 | −14.7 | −3.3 | 2.7 | 7.1 | −3.9 | 32.0 | 2.9 |
| **Chloride** | | | | | | | | | | | | | |
| Input | 5.5[b] | 4.5[b] | 2.6[d] | 6.2[b] | 5.0 | 6.4 | 5.3 | 11.1 | 5.5 | 11.4 | 4.4 | 67.9 | 6.2 |
| Output | 3.7[b] | 2.8[b] | 4.3 | 4.8 | 5.3 | 5.2 | 4.2 | 4.6 | 3.6 | 6.0 | 5.9 | 50.4 | 4.6 |
| Net | 1.8 | 1.7 | −1.7 | 1.4 | −0.3 | 1.2 | 1.1 | 6.5 | 1.9 | 5.4 | −1.5 | 17.5 | 1.6 |
| **Phosphate** | | | | | | | | | | | | | |
| Input | 0.09[b] | 0.08[b] | 0.10[b] | 0.11[b] | 0.11[b] | 0.10[b] | 0.10[b] | 0.10[b] | 0.10[b] | 0.14 | 0.13 | 1.16 | 0.11 |
| Output | 0.02[b] | 0.01[b] | 0.02[b] | 0.02[b] | 0.02[b] | 0.028 | 0.02[b] | 0.02[b] | 0.02[b] | 0.018 | 0.029 | 0.225 | 0.020 |
| Net | 0.07 | 0.07 | 0.08 | 0.09 | 0.09 | 0.07 | 0.08 | 0.08 | 0.08 | 0.12 | 0.10 | 0.94 | 0.09 |
| **Bicarbonate[e]** | | | | | | | | | | | | | |
| Input | a | a | a | a | a | a | a[a] | a | a | a | a | a | a |
| Output | 6.2[b] | 4.6[b] | 6.2 | 9.4 | 9.6 | 7.0 | 6.0 | 7.1[b] | 6.6[b] | 9.0[b] | 12.5[b] | 84.2 | 7.7 |
| Net | −6.2 | −4.6 | −6.2 | −9.4 | −9.6 | −7.0 | −6.0 | −7.1 | −6.6 | −9.0 | −12.5 | −84.2 | −7.7 |
| **Dissolved silica (SiO₂)** | | | | | | | | | | | | | |
| Input | a | a | a | a | a | a | a | a | a | a | a | a | a |
| Output | 30.5[b] | 20.8 | 36.1 | 41.6 | 42.1 | 35.0 | 41.4 | 33.8 | 30.6 | 45.7 | 56.7 | 414.3 | 37.7 |
| Net | −30.5 | −20.8 | −36.1 | −41.6 | −42.1 | −35.0 | −41.4 | −33.8 | −30.6 | −45.7 | −56.7 | −414.3 | −37.7 |

[a] Not measured, but trace quantities.

[b] Calculated value, based on weighted average concentration during years when chemical measurements were made and on amount of precipitation or streamflow during the specific year.

[c] Calculated from weighted average concentration for 1964–1966 times precipitation or streamflow for 1963–1964.

[d] Based on annual concentration of 0.50 mg/l (Juang and Johnson, 1967).

[e] Watershed 4 only.

**Table 11.** Mean Annual (Arithmetic Average of Annual Values) Bulk Precipitation Input, Gross Stream-Water Output and Net Loss or Gain, and Standard Deviation of the Mean for Dissolved Substances, during 1963–1974 for the Hubbard Brook Experimental Forest

| Substance | Input | | | | | Gross output | | | | | Net loss or gain[b] | |
|---|---|---|---|---|---|---|---|---|---|---|---|---|
| | kg/ha | % of total | Eq × $10^3$/ha | % of total | n[a] | kg/ha | % of total | Eq × $10^3$/ha | % of total | n[a] | kg/ha | Eq × $10^3$/ha |
| $Ca^{2+}$ | 2.17 ± 0.21 | 3 | 0.108 ± 0.010 | 8 | 11 | 13.7 ± 1.1 | 9 | 0.684 ± 0.055 | 42 | 11 | −11.6 ± 1.2 | −0.579 ± 0.060 |
| $Mg^{2+}$ | 0.58 ± 0.07 | 1 | 0.048 ± 0.006 | 3 | 11 | 3.13 ± 0.21 | 2 | 0.257 ± 0.017 | 16 | 11 | −2.55 ± 0.26 | −0.210 ± 0.021 |
| $K^+$ | 0.89 ± 0.19 | 1 | 0.023 ± 0.005 | 2 | 11 | 1.89 ± 0.14 | 1 | 0.048 ± 0.004 | 3 | 11 | −0.99 ± 0.27 | −0.025 ± 0.007 |
| $Na^+$ | 1.59 ± 0.12 | 2 | 0.069 ± 0.005 | 5 | 11 | 7.23 ± 0.47 | 5 | 0.315 ± 0.020 | 19 | 11 | −5.64 ± 0.50 | −0.245 ± 0.022 |
| $Al^{3+}$ | — | | — | — | — | 1.87 ± 0.16 | 1 | 0.208 ± 0.018 | 13 | 6 | −1.87 ± 0.16 | −0.208 ± 0.018 |
| $NH_4^+$ | 2.90 ± 0.18 | 4 | 0.161 ± 0.010 | 12 | 10 | 0.34 ± 0.08 | <1 | 0.019 ± 0.004 | 1 | 10 | +2.55 ± 0.21 | +0.141 ± 0.012 |
| $H^+$ | 0.977 ± 0.043 | 1 | 0.969 ± 0.043 | 70 | 10 | 0.114 ± 0.015 | <1 | 0.103 ± 0.015 | 6 | 10 | +0.872 ± 0.03 | +0.865 ± 0.033 |
| $SO_4^{2-}$ | 38.4 ± 2.5 | 52 | 0.799 ± 0.052 | 61 | 10 | 53.8 ± 4.4 | 36 | 1.12 ± 0.092 | 67 | 10 | −15.3 ± 2.8 | −0.319 ± 0.058 |
| $NO_3^-$ | 19.7 ± 2.1 | 27 | 0.317 ± 0.034 | 24 | 10 | 17.1 ± 3.2 | 11 | 0.275 ± 0.052 | 17 | 10 | +2.6 ± 2.7 | +0.042 ± 0.044 |
| $Cl^-$ | 7.01 ± 1.1 | 9 | 0.198 ± 0.032 | 15 | 7 | 4.89 ± 0.33 | 3 | 0.141 ± 0.009 | 8 | 7 | +2.02 ± 1.10 | +0.057 ± 0.031 |
| $PO_4^{3-}$ | 0.131 | <1 | 0.004 | <1 | 2 | 0.023 | <1 | <0.001 | <1 | 2 | +0.108 | +0.003 |
| Dissolved silica | — | | — | — | — | 38.4 ± 3.0 | 26 | — | — | 10 | −38.4 ± 3.0 | — |
| $HCO_3^-$ | — | | — | — | — | 7.64 ± 0.77 | 5 | 0.125 ± 0.013 | 8 | 5 | −7.64 ± 0.77 | −0.125 ± 0.013 |
| Total | 74.3 | 100 | 1.378 (cations) 1.318 (anions) | | | 150.1 | 100 | 1.634 (cations) 1.661 (anions) | | | −75.8 | −0.261 (cations) −0.342 (anions) |

[a]number of years of measurement.

[b]There are small rounding errors associated with some of these means and their calculation from gross output minus input.

these undisturbed aggrading watershed ecosystems each year (Tables 10 and 11). The long-term averages suggest that there has been a small net gain in nitrate and chloride each year but there is no statistical difference ($p > 0.05$) between the annual inputs in bulk precipitation and stream-water outputs (Table 11).

There are two separate considerations to evaluating annual input–output budgets: (1) The *direction* of the net change, whether input is greater or less than output; and (2) the *magnitude* of the net change, i.e., the difference in amount between input and output. These parameters may vary from year to year for the individual chemical elements. For some ions, data from a single annual cycle ($n = 1$) are sufficient to establish the *direction* of net change; for other ions data from several years ($n > 1$) are necessary. Similarly, a variable number of years are required to characterize the *magnitude* of net change with meaningful confidence limits. Based on 11 yr of study at Hubbard Brook it is possible to make some generalizations about the annual budgets for the various ions. Generally, the direction of net change for budgets of precipitation input minus stream-water output for $Ca^{2+}$, $Mg^{2+}$, $Na^+$, $Al^{3+}$, $NH_4^+$, $H^+$, $SO_4^{2-}$, $HCO_3^-$, and dissolved silica was predictable from only a few years (1–3) of data (Table 11). For these ions the direction of the budget is not in doubt, but the magnitude may be. For $K^+$, $NO_3^-$, and $Cl^-$ one or a few years were liable to give unreliable results for both net change and magnitude. With the coefficient of variability (standard deviation of the mean divided by the mean net change; Table 11) used as a guide, a few annual measurements would yield reliable estimates of magnitude of loss or gain for $Ca^{2+}$, $Mg^{2+}$, $Na^+$, $Al^{3+}$, $NH_4^+$, $H^+$, $SiO_2$, and $HCO_3^-$. Nitrate, $Cl^-$, and $K^+$ also have a large range in annual magnitude of net change and they are much more unpredictable relative to direction of net change than the other ions (Tables 10 and 11). Annual precipitation inputs for potassium and chloride have been particularly variable (Table 11). The standard deviation of the mean for long-term potassium inputs is 22%, and for chloride it is 17% of the mean. These highly variable inputs are therefore the major reason for the uncertainty in the annual budgets.

These data emphasize that for some elements data from one or a few annual cycles may give grossly aberrant estimates of long-term mean values of both direction and magnitude of change.

## Sources of input other than bulk precipitation

*Nitrogen.* During the 11-yr period 1963–1974, 209.5 kg/ha of nitrate (47.3 kg/ha $NO_3$-N) were added in bulk precipitation and 177.5 kg of $NO_3^-$ per hectare (40.1 kg/ha $NO_3$-N) were lost in stream water from these undisturbed watershed ecosystems at Hubbard Brook. Likewise, 31.6 kg of ammonium per hectare (24.6 kg/ha $NH_4$-N) have been added in precipitation and 3.70 kg of $NH_4^+$ per hectare (2.9 kg/ha $NH_4$-N) have been lost in stream water during 1963–1974. This gives a total dissolved inorganic nitrogen input of 71.9 kg/ha and a nitrogen output of 43.0 kg/ha during this 11-yr period. These long-term data show that more inorganic nitrogen has been added to these forested ecosystems in precipitation than has been lost in stream water. This amount, although substantial, cannot account for nitrogen accumulating within the ecosystem in living and dead biomass (Bormann et al., 1977; Covington, 1975, 1976; Whittaker et al., 1974). Nitrogen is incorporated in this biomass and thereby is stored at a rate of about 16.7 kg/ha·yr in these undisturbed ecosystems. Because nitrogen is essentially absent from the bedrock and till at Hubbard Brook, we assume that the release of nitrogenous substances by weathering and the formation of secondary nitrogenous minerals are negligible. Therefore, the net gaseous and aerosol exchange of nitrogen across the ecosystem's boundaries may be estimated (Figure 1) from net biomass accumulation minus net flux (precipitation input minus stream-water output). The long-term net gain in inorganic nitrogen is 2.5 kg/ha·yr (precipitation input minus loss of dissolved and particulate nitrogen compounds in streamflow; Tables 9–11). The net accumulation in living and dead biomass is 16.7 kg/ha·yr. Therefore, ecosystem analysis using the small watershed technique shows that on the average about 14.2 kg of nitrogen per hectare·year must enter the ecosystem in a gaseous form or from impacted nitrogenous aerosols. Volatile exchanges of gases with the atmosphere have not been fully evaluated for the ecosystems at Hubbard Brook. Preliminary field data indicate that microbial fixation of nitrogen occurs within these forested ecosystems (Roskoski, 1975) and that denitrification, particularly chemical denitrification, may occur. It is likely that much of the additional accumulation of nitrogen within the ecosystem is largely a result of fixation of gaseous $N_2$ by microorganisms. If denitrification produced significant losses of gaseous nitrogen, then the actual amount of $N_2$ fixed would be even greater. Consequently, the actual input component for N (Tables 10 and 11) should be increased by some 14.2 kg/ha·yr, thereby increasing the average net gain value for N by a similar amount (Table 12).

*Chlorine.* With the exception of the relatively very large input of chloride in precipitation during 1970–1971 and 1972–1973, the long-term annual budget for chloride is nearly in balance. As mentioned above, long-term annual inputs are not statistically different than long-term outputs (Table 11). The high chloride input in 1970–1971 was largely a result of three or four exceptionally high concentrations in autumnal rainfall samples. These samples showed no evidence of contamination and repeated chemical analysis verified the results. Generally, the chloride budget at Hubbard Brook might be expected to be in balance (Juang and Johnson, 1967). We have no quantitative data on the biologic accumulation of chloride in these ecosystems but it is presumed to be small. However, substantial releases of chloride from some internal reser-

**Table 12.** Summary of Annual Input–Output Budgets for Undisturbed Forested Ecosystems at Hubbard Brook[a]

| Element | Input, kg/ha·yr | Output, kg/ha·yr | Net gain (+) or loss (−), kg/ha·yr |
|---|---|---|---|
| Si | — | 23.8 | −23.8 |
| Ca | 2.2 | 13.9 | −11.7 |
| Na | 1.6 | 7.5 | −5.9 |
| Al | — | 3.4 | −3.4 |
| Mg | 0.6 | 3.3 | −2.7 |
| K | 0.9 | 2.4 | −1.5 |
| Organic C | 1484[b] | 12.3 | +1472 |
| N | 20.7 | 4.0 | +16.7 |
| Cl | 6.2 | 4.6 | +1.6 |
| S | 18.8 | 17.6 | +1.2 |
| H | 0.96[c] | 0.10[c] | +0.86 |
| P | 0.036 | 0.019 | +0.017 |

[a]Input values for organic C, N, and S include bulk precipitation, estimates of aerosol impaction, and net gaseous uptake; all others are based on bulk precipitation only. Output values include both dissolved substances (Table 10) and particulate matter (Table 9) in stream water.

[b]Includes ecosystem biomass accretion (*net* ecosystem gaseous uptake of $CO_2$) based on the period 1961–1965 (Whittaker, et al., 1974); therefore output does not include respiration losses of $CO_2$.

[c]Dissolved form only.

voir after disturbance by deforestation (Likens et al., 1970) suggest that chloride may be accumulating in the undisturbed forested ecosystem. Chlorine, like nitrogen, may have a gaseous phase at normal biologic temperatures; however, chlorine gas is very reactive and is not common in the atmosphere. Significant amounts of chloride in throughfall and stemflow (Eaton et al., 1973; Table 6) suggested that small amounts of impacted aerosols containing chloride were washed off vegetation surfaces or that chloride was leached from vegetation by incident precipitation. Unfortunately, our data do not provide any definitive answers in this regard.

*Sulfur.* Sulfur also may exist as a gas or as an aerosol in the atmosphere. Storage of sulfur in the annual biomass increment occurs at a rate of about 2.0 kg/ha·yr. A small amount is released from chemical weathering (0.8 kg/ha·yr) within the ecosystems at Hubbard Brook, but the algebraic sum of the long-term precipitation input, streamwater output, weathering, and biomass accumulation shows that some 6.1 kg/ha·yr of sulfur must be accumulated each year from atmospheric gaseous and aerosol sources. This amount should be added to the bulk precipitation input value (Tables 10 and 11) for these ecosystems (Table 12). The origin, quantification, and ecological effects of gaseous or aerosol sulfur require detailed studies to elucidate the role of sulfur in forested ecosystems. Heretofore, these fluxes have been ignored or have not been quantitatively evaluated in biogeochemical studies.

## Input–output budgets: summary

Overall, during the period 1963–1974 there was an annual net loss of total dissolved inorganic substances from the experimental watersheds amounting to 74.7 kg/ha·yr (Table 10). The net loss averaged 261 Eq/ha·yr for cations and 342 Eq/ha·yr for anions (Table 11). The average net output of dissolved inorganic substances minus dissolved silica (1963–1974) was 37.1 kg/ha·yr. However, sizeable yearly variations have occurred during the study (Figures 14 and 20; Tables 10 and 11). For example, the smallest net loss of dissolved inorganic substances (27.8 kg/ha, or 7.0 kg/ha for total material minus dissolved silica) occurred during 1964–1965, the driest year of our study. The largest net losses of dissolved inorganic substances occurred during the very wet year, 1973–1974 (139.7 kg/ha and 986 cationic equivalents per hectare). Budgets adjusted to all sources of input are presented in Table 12. The explanation for the overall net losses lies in the biogeochemical reactions that occur within the ecosystem, notably chemical weathering and related phenomena.

**Seasonal variations** Climate and biologic activity at Hubbard Brook are distinctly seasonal and much of the biogeochemistry reflects these climatic patterns. Normally the snowpack persists from about mid-November until April (Table 2). The trees start to produce leaves during late May, and the major leaf fall occurs in early October. Some organisms of the ecosystem have adapted to utilize efficiently the environmental conditions of the transition periods between seasons (cf. Muller and Bormann, 1976).

We have divided the year into four seasons to investigate the short-term variations in chemistry of precipitation and stream water. Summer conditions are assumed from 1 June to 30 September, autumn from 1 October to 30 November, winter from 1 December to 28 February, and spring from 1 March to 31 May. Although these periods are arbitrary, they correspond well with actual conditions and biologic activity at Hubbard Brook.

The weighted concentrations of various ions in precipitation are highly variable on a storm or weekly basis but some seasonal trends have emerged. The total ionic concentration of precipitation during the winter at Hubbard Brook is almost one-half that of other seasons of the year (Table 13). This is reflected in the concentration of individ-

**Table 13.** Weighted Average Concentrations in Bulk Precipitation During Different Seasonal Periods for the Hubbard Brook Experimental Forest

| Substance | No. of years | June–Sept. mg/l | June–Sept. $\mu$Eq/l | Oct.–Nov. mg/l | Oct.–Nov. $\mu$Eq/l | Dec.–Feb. mg/l | Dec.–Feb. $\mu$Eq/l | March–May mg/l | March–May $\mu$Eq/l |
|---|---|---|---|---|---|---|---|---|---|
| $Ca^{2+}$ | 11 | 0.14 | 6.99 | 0.18 | 8.98 | 0.11 | 5.49 | 0.24 | 12.0 |
| $Mg^{2+}$ | 11 | 0.04 | 3.29 | 0.07 | 5.76 | 0.03 | 2.47 | 0.05 | 4.11 |
| $K^+$ | 11 | 0.07 | 1.79 | 0.11 | 2.81 | 0.04 | 1.02 | 0.07 | 1.79 |
| $Na^+$ | 11 | 0.09 | 3.91 | 0.19 | 8.26 | 0.11 | 4.78 | 0.12 | 5.22 |
| $NH_4^+$ | 10 | 0.24 | 13.3 | 0.19 | 10.5 | 0.15 | 8.32 | 0.27 | 15.0 |
| $H^+$ | 10 | 0.090 | 89.3 | 0.070 | 69.4 | 0.045 | 44.6 | 0.079 | 78.4 |
| $SO_4^{2-}$ | 10 | 3.72 | 77.5 | 2.34 | 48.7 | 1.67 | 34.8 | 3.23 | 67.2 |
| $NO_3^-$ | 10 | 1.32 | 21.3 | 1.47 | 23.7 | 1.39 | 22.4 | 1.77 | 28.6 |
| $Cl^-$ | 7 | 0.50 | 14.1 | 1.04 | 29.3 | 0.30 | 8.46 | 0.36 | 10.2 |
| $PO_4^{3-}$ | 2 | 0.008 | 0.25 | 0.008 | 0.25 | 0.004 | 0.13 | 0.010 | 0.32 |
| Total | | 6.22 | | 5.67 | | 3.85 | | 6.20 | |
| $\Sigma^+$ | | | 118.6 | | 105.7 | | 66.7 | | 116.5 |
| $\Sigma^-$ | | | 112.9 | | 101.7 | | 65.7 | | 106.0 |

ual ions, except for sodium and nitrate. Concentrations of $Na^+$ and $NO_3^-$ during the winter are about equal to or greater than their concentrations during other seasons. A peak in concentration for chloride, sodium, potassium, and magnesium occurs in the autumn (October; Table 14). This annual sinusoidal pattern for many of the elements may result in part from the origin of air masses and in part from differences in aerosol scavenging efficiency between raindrops and snowflakes. Herman and Gorham (1957) have suggested from their studies in Nova Scotia that snowflakes may be less efficient than raindrops in removing materials from the atmosphere. Nitrate concentrations in precipitation are relatively similar during all seasons (Table 13). Other obvious but unexplained variations include: (1) Calcium equivalent concentrations are much greater than (almost double) sodium values during the spring and summer but about equal to sodium during the autumn and winter; and (2) the equivalent concentration of chloride exceeds that of nitrate during autumn but nitrate concentrations greatly exceed chloride values during the other seasons. The relatively high concentrations of sodium and chloride in autumn rains may reflect a larger influence from marine aerosols at this time in the eastern United States. However, chloride : sodium ratios during summer and autumn are identical (5.6 on a weight basis) and are far in excess of those expected (1.8) from seawater (Junge, 1963). The high Cl : Na ratios during summer and autumn at Hubbard Brook may indicate some seasonal transport of the chloride to the area from distant anthropogenic sources (Gorham, 1958; Junge, 1963).

Concentrations of various ions in stream water are much less variable seasonally than in precipitation (Table 15). Concentrations in stream water are slightly lower in the summer, primarily because of the intense biologic utilization of potassium and nitrate during the growing season.

In general, nitrate concentrations in stream water begin to rise in the autumn when there is a reduction in biologic activity, and attain a maximum concentration in the winter or early spring. United States Geological Survey Water Resources data show a similar pattern for rivers and streams throughout New England and New York State. This pattern also has been observed in the surface waters of many freshwater lakes where nitrate concentrations reach maximum levels in winter and early spring. This latter increase in nitrate concentrations is attributed to

**Table 14.** Monthly Bulk Precipitation and Stream-Water Values for the Hubbard Brook Experimental Forest during 1963–1974[a]

| | | Water[b] | Ca²⁺[b] | Mg²⁺[b] | K⁺[b] | Na⁺[b] | H⁺[c] | NH₄⁺[c] | NO₃⁻[c] | SO₄²⁻[c] | Cl⁻[d] | PO₄³⁻[e] |
|---|---|---|---|---|---|---|---|---|---|---|---|---|
| **Bulk precipitation** | | | | | | | | | | | | |
| Summer | June | 12.07 | 0.15 | 0.05 | 0.10 | 0.13 | 0.0787 | 0.23 | 1.44 | 3.8 | 0.83 | 0.007 |
| | July | 11.07 | 0.17 | 0.05 | 0.07 | 0.07 | 0.1054 | 0.27 | 1.46 | 4.4 | 0.25 | 0.010 |
| | August | 13.38 | 0.13 | 0.03 | 0.04 | 0.06 | 0.0751 | 0.23 | 1.24 | 3.2 | 0.39 | 0.012 |
| Autumn | September | 9.56 | 0.12 | 0.04 | 0.06 | 0.08 | 0.1041 | 0.23 | 1.53 | 3.5 | 0.44 | 0.007 |
| | October | 8.63 | 0.25 | 0.08 | 0.13 | 0.23 | 0.0786 | 0.23 | 1.69 | 2.6 | 1.00 | 0.013 |
| | November | 13.58 | 0.14 | 0.07 | 0.09 | 0.17 | 0.0644 | 0.16 | 1.54 | 2.2 | 1.07 | 0.005 |
| Winter | December | 14.93 | 0.09 | 0.03 | 0.05 | 0.09 | 0.0432 | 0.11 | 1.37 | 1.6 | 0.30 | 0.005 |
| | January | 7.96 | 0.10 | 0.02 | 0.04 | 0.13 | 0.0528 | 0.23 | 1.89 | 2.1 | 0.29 | 0.005 |
| | February | 9.29 | 0.15 | 0.04 | 0.03 | 0.14 | 0.0423 | 0.15 | 1.37 | 1.4 | 0.32 | 0.001 |
| Spring | March | 10.25 | 0.20 | 0.04 | 0.04 | 0.12 | 0.0630 | 0.21 | 1.94 | 2.5 | 0.33 | 0.003 |
| | April | 9.86 | 0.27 | 0.06 | 0.06 | 0.11 | 0.0779 | 0.31 | 1.88 | 3.1 | 0.31 | 0.013 |
| | May | 11.67 | 0.25 | 0.05 | 0.10 | 0.11 | 0.0935 | 0.30 | 1.88 | 3.9 | 0.42 | 0.013 |
| **Stream water** | | | | | | | | | | | | |
| Summer | June | 4.91 | 1.46 | 0.32 | 0.16 | 0.84 | 0.0127[f] | 0.02 | 0.54 | 6.6 | 0.38 | 0.002 |
| | July | 2.28 | 1.54 | 0.33 | 0.14 | 0.87 | 0.0135[f] | 0.04 | 0.45 | 6.8 | 0.37 | 0.002 |
| | August | 1.84 | 1.45 | 0.33 | 0.13 | 0.91 | 0.0099[f] | 0.08 | 0.35 | 6.3 | 0.43 | 0.004 |
| Autumn | September | 1.58 | 1.57 | 0.40 | 0.11 | 1.17 | 0.0082[f] | 0.05 | 0.27 | 7.4 | 0.57 | 0.003 |
| | October | 3.83 | 1.58 | 0.40 | 0.22 | 1.01 | 0.0086[f] | 0.06 | 0.31 | 7.0 | 0.56 | 0.002 |
| | November | 7.92 | 1.74 | 0.41 | 0.23 | 0.97 | 0.0098[f] | 0.09 | 1.20 | 6.8 | 0.70 | 0.001 |
| Winter | December | 8.11 | 1.71 | 0.39 | 0.20 | 0.89 | 0.0131[f] | 0.04 | 2.46 | 6.3 | 0.49 | 0.002 |
| | January | 3.86 | 1.69 | 0.39 | 0.20 | 0.96 | 0.0118[f] | 0.03 | 2.17 | 6.4 | 0.54 | 0.002 |
| | February | 3.66 | 1.74 | 0.41 | 0.21 | 0.98 | 0.0107[f] | 0.05 | 2.72 | 6.2 | 0.51 | 0.001 |
| Spring | March | 9.32 | 1.71 | 0.40 | 0.24 | 0.92 | 0.0115[f] | 0.04 | 2.80 | 6.3 | 0.54 | 0.001 |
| | April | 23.85 | 1.69 | 0.38 | 0.26 | 0.77 | 0.0135[f] | 0.03 | 2.95 | 6.1 | 0.51 | 0.002 |
| | May | 12.11 | 1.56 | 0.34 | 0.25 | 0.79 | 0.0143[f] | 0.02 | 1.67 | 6.2 | 0.46 | 0.002 |

[a]Mean weighted concentrations of dissolved substances given in mg/l; water given as cm/unit area.

[b]1963–1974.

[c]1964–1974.

[d]1967–1974.

[e]1972–1974.

[f]1965–1974.

**Table 15.**  Weighted Average Concentrations in Stream Water during Different Seasonal Periods for Undisturbed Watersheds of the Hubbard Brook Experimental Forest

| Substance | No. of years | June–Sept. | | Oct.–Nov. | | Dec.–Feb. | | March–May | |
|---|---|---|---|---|---|---|---|---|---|
| | | mg/l | μEq/l | mg/l | μEq/l | mg/l | μEq/l | mg/l | μEq/l |
| $Ca^{2+}$ | 11 | 1.49 | 74.4 | 1.69 | 84.3 | 1.71 | 85.3 | 1.66 | 82.8 |
| $Mg^{2+}$ | 11 | 0.34 | 28.0 | 0.41 | 33.7 | 0.40 | 32.9 | 0.37 | 30.4 |
| $K^+$ | 11 | 0.14 | 3.58 | 0.23 | 5.88 | 0.20 | 5.11 | 0.25 | 6.39 |
| $Na^+$ | 11 | 0.91 | 39.6 | 0.98 | 42.6 | 0.92 | 40.0 | 0.81 | 35.2 |
| $NH_4^+$ | 10 | 0.04 | 2.22 | 0.08 | 4.44 | 0.04 | 2.22 | 0.03 | 1.66 |
| $H^+$ | 9 | 0.012 | 11.9 | 0.009 | 8.93 | 0.012 | 11.9 | 0.013 | 12.9 |
| $SO_4^{2-}$ | 10 | 6.7 | 139 | 6.8 | 142 | 6.3 | 131 | 6.2 | 129 |
| $NO_3^-$ | 10 | 0.44 | 7.10 | 0.89 | 14.4 | 2.46 | 39.7 | 2.57 | 41.5 |
| $Cl^-$ | 10 | 0.49 | 13.8 | 0.62 | 17.5 | 0.57 | 16.1 | 0.54 | 15.2 |
| $PO_4^{3-}$ | 2 | 0.003 | 0.09 | 0.001 | 0.03 | 0.002 | 0.06 | 0.002 | 0.06 |
| Total | | 10.6 | | 11.7 | | 12.6 | | 12.4 | |
| $\Sigma^+$ | | | 158.7 | | 179.8 | | 177.4 | | 169.3 |
| $\Sigma^-$ | | | 160.0 | | 173.9 | | 186.9 | | 185.8 |

increased nitrification during the winter (Hutchinson 1957, p. 871)

At Hubbard Brook we have observed that nitrate concentrations increase markedly with increased streamflow (Johnson et al. 1969). However this may be a fortuitous relationship since the late winter-early spring period of maximum streamflow is potentially also a time for relatively high nitrification in the soil. Even though there may be no causal relationship between high nitrate concentration and increased streamflow during this time, the fact that this period of high nitrate concentrations also is characterized by large discharge remains of prime importance to nutrient budget calculations.

The seasonal pattern for ammonium concentrations in stream water is somewhat unique. The clear peak in ammonium concentration during the autumn may reflect increased decomposition of organic matter in the forest floor after leaf fall.

The pattern of average precipitation input and streamwater output during the seasonal periods (Tables 16 and 17) is similar to the monthly flux (Figures 24–28), but the resolution is not as great. Input (on a weight basis) is greatest during the summer and least during the autumn (Table 16). The equivalent input is lowest during the winter, largely because of the sharp decline in chloride inputs. On a weight basis, sulfate and nitrate provide about 70–80% of the dissolved inorganic substances in precipitation.

**Table 16.** Average Precipitation Input during Different Seasonal Periods for Undisturbed Watersheds of the Hubbard Brook Experimental Forest

| Substance | No. of years | June–September | | | October–November | | | December–February | | | March–May | | |
|---|---|---|---|---|---|---|---|---|---|---|---|---|---|
| | | g/ha | Eq/ha | % of annual total | g/ha | Eq/ha | % of annual total | g/ha | Eq/ha | % of annual total | g/ha | Eq/ha | % of annual total |
| $Ca^{2+}$ | 11 | 656 | 32.7 | 30.2 | 405 | 20.2 | 18.7 | 343 | 17.1 | 15.8 | 767 | 38.3 | 35.3 |
| $Mg^{2+}$ | 11 | 182 | 15.0 | 31.3 | 153 | 12.6 | 26.4 | 89.6 | 7.37 | 15.4 | 156 | 12.8 | 26.9 |
| $K^+$ | 11 | 310 | 7.93 | 34.7 | 238 | 6.09 | 26.6 | 126 | 3.22 | 14.1 | 220 | 5.63 | 24.6 |
| $Na^+$ | 11 | 393 | 17.1 | 25.3 | 431 | 18.7 | 27.7 | 361 | 15.7 | 23.2 | 369 | 16.1 | 23.7 |
| $NH_4^+$ | 10 | 1,140 | 63.2 | 39.3 | 413 | 22.9 | 14.2 | 491 | 27.2 | 16.9 | 859 | 47.6 | 29.6 |
| $H^+$ | 10 | 425 | 422 | 43.5 | 155 | 154 | 15.9 | 149 | 148 | 15.3 | 248 | 246 | 25.4 |
| $SO_4^{2-}$ | 10 | 17,600 | 366 | 45.9 | 5,170 | 108 | 13.5 | 5,500 | 115 | 14.3 | 10,100 | 210 | 26.3 |
| $NO_3^-$ | 10 | 6,270 | 101 | 31.9 | 3,240 | 52.3 | 16.5 | 4,580 | 73.9 | 23.3 | 5,550 | 89.5 | 28.3 |
| $Cl^-$ | 7 | 2,380 | 67.1 | 34.0 | 2,290 | 64.6 | 32.7 | 1,110 | 31.3 | 15.8 | 1,230 | 34.7 | 17.5 |
| $PO_4^{3-}$ | 2 | 51.1 | 1.61 | 39.0 | 20.0 | 0.632 | 15.3 | 19.1 | 0.603 | 14.6 | 40.7 | 1.29 | 31.1 |
| Total | — | 29,356 | $\Sigma + 558$ $\Sigma - 536$ | 39.5 | 12,515 | $\Sigma + 234$ $\Sigma - 226$ | 16.8 | 12,769 | $\Sigma + 219$ $\Sigma - 221$ | 17.2 | 19,540 | $\Sigma + 366$ $\Sigma - 335$ | 26.3 |
| Water (cm/ha) | 11 | 46.0 | | 34.8 | 22.2 | | 16.8 | 32.2 | | 24.4 | 31.8 | | 24.1 |

**Table 17.** Average Stream-Water Output during Different Seasonal Periods for Undisturbed Watersheds of the Hubbard Brook Experimental Forest

| Substance | No. of years | June–September | | | October–November | | | December–February | | | March–May | | |
|---|---|---|---|---|---|---|---|---|---|---|---|---|---|
| | | g/ha | Eq/ha | % of annual total | g/ha | Eq/ha | % of annual total | g/ha | Eq/ha | % of annual total | g/ha | Eq/ha | % of annual total |
| $Ca^{2+}$ | 11 | 1,580 | 78.8 | 11.5 | 1,980 | 98.8 | 14.4 | 2,680 | 134 | 19.5 | 7,510 | 375 | 54.6 |
| $Mg^{2+}$ | 11 | 357 | 29.4 | 11.4 | 476 | 39.2 | 15.1 | 619 | 50.9 | 19.7 | 1,690 | 139 | 53.8 |
| $K^+$ | 11 | 151 | 3.86 | 8.0 | 268 | 6.85 | 14.2 | 316 | 8.08 | 16.8 | 1,150 | 29.4 | 61.0 |
| $Na^+$ | 11 | 965 | 42.0 | 13.3 | 1,160 | 50.5 | 16.0 | 1,450 | 63.1 | 20.0 | 3,660 | 159 | 40.6 |
| $NH_4^+$ | 10 | 45.0 | 2.49 | 13.4 | 96.6 | 5.36 | 28.7 | 57.4 | 3.18 | 17.0 | 138 | 7.65 | 40.9 |
| $H^+$ | 9 | 14.4 | 14.3 | 13.2 | 11.9 | 11.8 | 10.9 | 19.6 | 19.4 | 17.9 | 63.4 | 629 | 58.0 |
| $SO_4^{2-}$ | 10 | 7,740 | 161 | 14.3 | 8,240 | 172 | 15.3 | 9,940 | 207 | 18.4 | 28,100 | 585 | 52.0 |
| $NO_3^-$ | 10 | 513 | 8.27 | 3.0 | 1,070 | 17.3 | 6.2 | 3,880 | 62.6 | 22.6 | 11,700 | 189 | 68.2 |
| $Cl^-$ | 10 | 569 | 16.1 | 12.2 | 746 | 21.0 | 15.9 | 903 | 25.5 | 19.3 | 2,460 | 69.4 | 52.6 |
| $PO_4^{3-}$ | 2 | 6.72 | 0.212 | 29.1 | 1.72 | 0.054 | 7.4 | 5.80 | 0.183 | 25.1 | 889 | 0.281 | 38.4 |
| Total | — | 11,941 | Σ + 171 Σ − 186 | 11.7 | 14,050 | Σ + 213 Σ − 210 | 13.7 | 19,871 | Σ + 279 Σ − 295 | 19.4 | 56,480 | Σ + 773 Σ − 843 | 55.2 |
| Water (cm/ha) | 11 | 10.6 | | 12.7 | 11.7 | | 14.1 | 15.6 | | 18.7 | 45.3 | | 54.4 |

On an equivalent basis, hydrogen ion and sulfate dominate, as expected. In general the input of nitrate reflects
more closely than any other ion the amount of water added
as precipitation (Table 16). Inputs of magnesium, potassium, and to a lesser extent calcium and sodium are relatively greater than expected on the basis of amount of
precipitation during the autumn (enriched precipitation
relative to the annual average) and less during the winter.
All ionic inputs during the winter season were less than
expected on the basis of amount of precipitation; in general, however, inputs of individual dissolved substances
were highly variable on a seasonal basis (Table 16).

More than 55% of the annual output of dissolved substances in stream water occurs during the spring. In contrast, only 11% of the output occurs during summer (Table
17). Sulfate clearly dominates the output of dissolved substances on a weight and equivalent basis. Calcium, magnesium, sulfate, chloride, and to a lesser extent hydrogen ion
and sodium losses in stream water reflect the amount of
water lost. Phosphate losses bear little relationship to the
amount of streamflow; summer and winter losses are
appreciably greater than expected and autumn and spring
losses are less than expected from the amount of streamflow. Likewise, output of nitrate during summer and
autumn is less than expected and during winter and spring
is more than expected on the basis of water loss. Outputs
of potassium and ammonium are variable, but the reduction of potassium during the summer produces an obvious
drop in stream-water output assumed to be from biologic
utilization within the forested ecosystem (Table 17).

**Monthly variations**     A greater insight into the biogeochemistry of these undisturbed ecosystems can be gained by looking at the nutrient
budgets on a monthly scale. Three patterns of input–
output emerge: those in which input exceeds output in
every month, those in which output exceeds input in every
month, and those in which there is a crossover from input
dominance to output dominance on a monthly basis. These
patterns are not related to whether the elements have a
gaseous component or are entirely sedimentary. For
example, Ca, P, and K are all sedimentary, yet they
exhibit different patterns; S and N, however, both atmos-
philic, exhibit two patterns. Instead of this simple assumption, a complex interaction involving such factors as input,
effect of biologic activity, and annual variations in weather
are determining factors.

**FIGURE 24.** Monthly flux of calcium, magnesium, and sodium for undisturbed watershed ecosystems of the Hubbard Brook Experimental Forest. Vertical bars indicate ± one standard deviation of the mean. Solid line is input and dashed line is output.

### Output consistently greater than input

Magnesium exemplifies a nutrient in which the output exceeds the input each month of the year on a long-term basis (Figure 24). The effects of increased streamflow are clearly shown by the peak chemical ouputs in April and November–December. During a specific year, the monthly input and output may be quite variable, and monthly outputs may not exceed inputs. In fact, based on a 6-yr average (1963–1969) inputs exceeded the outputs during July. Calcium, sodium, and aluminum all have patterns similar to magnesium; i.e., outputs consistently exceed inputs on a monthly basis (Figure 24).

### Input consistently greater than output

In sharp contrast to magnesium, ammonium is characterized by monthly inputs that consistently exceed outputs on a long-term basis (Figure 25). Monthly outputs of ammonium are relatively constant and low. The pattern of ammonium inputs may be highly variable for a specific year but it becomes much clearer and distinctly sinusoidal with long-term data. The monthly input during late autumn and winter is relatively low. This may reflect a reduced rate of ammonia input into the atmosphere from decay in

**FIGURE 25.** Monthly flux of ammonium for undisturbed watershed ecosystems of the Hubbard Brook Experimental Forest. Vertical bars indicate ± one standard deviation of the mean. Solid line is input and dashed line is output.

the soil because of the lower temperatures, or the transfer of ammonia to the atmosphere may be reduced by the presence of a snowpack. Likewise, higher inputs during the spring and summer may reflect the general increase in biologic activity of the soil and the elevated generation of ammonia. However, because ammonia may be converted rapidly to nitrate, $N_2O$, $N_2$, etc., it is difficult to interpret these monthly fluxes. The overall monthly fluxes for hydrogen and phosphate are similar to ammonium (input exceeds output). However we do not have long-term data on phosphorus budgets.

### Crossover patterns

Potassium inputs exceed outputs from mid-May until mid-October and during the remainder of the year monthly outputs are larger than inputs (Figure 26). This is of interest, for we have suggested earlier that the potassium budget (particularly potassium in stream water) is particularly sensitive to seasonal biologic control within the ecosystem (Tables 15 and 17). As mentioned above, the forest vegetation of these watersheds normally breaks dormancy about mid-May and the deciduous leaves are shed about mid-October. Apparently the biotic portion of the ecosystem actively extracts potassium from the soil and drainage waters during the growing season (cf. Johnson et al., 1969). The "crossover times" separating input or output domination were not clearly defined in the "long-term" monthly budgets by 1969 (1963–1969), which presumably was a residual effect of the drought years (1963–1965) on the cumulative average (see Table 10). The large monthly export of potassium in stream water during March through May of each year generally surpasses the average annual input to these undisturbed ecosystems (Figure 26; Tables 16 and 17). The monthly pattern of potassium inputs and outputs for any specific year may be highly variable. For example, as mentioned above, inputs exceeded outputs during the drought years of 1963–1965 (Figure 27) but less dramatic, although complicated, patterns occurred in other years.

The monthly pattern for sulfate is generally similar to that of potassium (Figure 26) except that the long-term results show that the crossover from input to output dominance in the autumn occurs somewhat earlier (late September) than for potassium. The monthly input of sulfate

**FIGURE 26.** Monthly flux of sulfate and potassium for undisturbed watershed ecosystems of the Hubbard Brook Experimental Forest. Vertical bars indicate ± one standard deviation of the mean. Solid line is input and dashed line is output.

**FIGURE 27.** Comparison of the monthly flux of potassium during the drought years of 1963–1965 with the long-term (1963–1974) average. Vertical bars indicate ± one standard deviation of the mean. Solid line is input and dashed line is output.

in bulk precipitation has a rather symmetrical, sinusoidal pattern, with a peak in July and a low in January–February. During the summer and early autumn of 1968 the input was unusually small and during April the output was unusually large, both of which combined to produce the largest annual net output for any year except the very wet year of 1973–1974 (Table 10).

Monthly nitrate flux requires special consideration. The monthly pattern is generally consistent but distinctly different than the pattern for ammonium (Figures 25 and 28),

**FIGURE 28.** Monthly flux of nitrate and chloride for the undisturbed watershed ecosystems of the Hubbard Brook Experimental Forest. Vertical bars indicate ± one standard deviation of the mean. Solid line is input and dashed line is output.

although the annual patterns for these two nitrogenous ions are similar (i.e., inputs exceed outputs). The long-term pattern for nitrate after 5 yr (1964–1969) and after 8 yr (1964–1972) were similar and showed that inputs exceeded outputs during every month except April, even though after 8 yr the average output for April was 55% higher than it was after only 5 yr. However, during 1969–1970, 1970–1971, and 1973–1974 a distinct change in the annual pattern occurred (Table 10). During these 3 yr the long-term annual relationship, where input exceeds output, was reversed. The input during August, September, October, November, and January 1969–1970 was less than the long-term average but it was the relatively large outputs in April and to a lesser extent in December and February that overwhelmed the long-term annual budget pattern (Figure 28). As mentioned previously this change in pattern during 1969–1970 and 1973–1974 may have been related to the unusual occurrence of soil frost.

The trend of increasing concentration (and input) of nitrate in precipitation during the period of study (Figures 8 and 16; Table 10) also has affected the long-term monthly flux. As a result of these perturbations in precipitation and stream-water flux, the resolution of the "average" long-term monthly pattern for nitrate is not good. Nevertheless, based on 10 yr of data, monthly inputs of nitrate exceed outputs during all months except December, March, April, and May. Inputs and outputs are essentially balanced during December and May (Figure 28).

The long-term monthly pattern for chloride is similar to that for nitrate (Figure 28).

# 5 Weathering

Both the qualitative and the quantitative changes in water chemistry elicited during the passage of water through the ecosystem are related in part to the process of chemical weathering. The bulk ionic composition of water entering the Hubbard Brook ecosystem is essentially characterized by acid salts, such as $H_2SO_4$, $HNO_3$, and HCl. In contrast, water leaving the system is characterized mainly by neutral salts, such as $CaSO_4$, $Na_2SO_4$, $Mg(NO_3)_2$, and to a lesser extent by chlorides and bicarbonates. This qualitative change in chemistry exemplifies the general chemical weathering reactions. This reaction and some of the chemical pathways in an open system, such as the Hubbard Brook watershed ecosystem, are shown in Figure 29.

As $H^+$ is fixed within the system, basic cations ($M^+$) may be leached from the systems and primary silicate minerals ($M^+X$) are transformed into clays or hydroxides and left behind as a residuum (Figure 29). When mechanical erosion or other disturbance is sufficiently slow the mineral residuum interacts with organic substances to form soil.

The rate at which the chemical weathering reaction proceeds may be seen as the rate at which $H^+$ is supplied to the system (see Figure 29). An alternate viewpoint, however, for determining weathering rate is from the rate at which cations are withdrawn from the system, i.e., cationic denudation (Reynolds and Johnson, 1972).

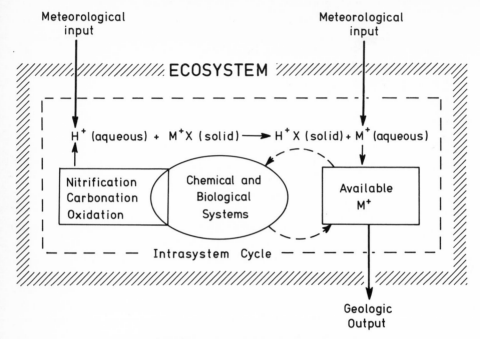

**FIGURE 29.** $M^+$ represents a basic cation and X is an ionic exchange substrate, usually a primary silicate mineral, but also includes such derivatives as soil minerals or soil humates. The above illustrates that as $H^+$ is added to the system from external and internal sources, basic cations ($M^+$) are leached from the system, and primary silicate minerals ($M^+X$) are transformed into secondary minerals ($H^+X$).

**Sources of hydrogen ion**   The hydrogen ion, which drives the chemical weathering reaction, is supplied from two sources at Hubbard Brook, one external to the ecosystem and the other internal. The external source is acids supplied in bulk precipitation per se (meteorologic input); the internal source is various biologic and chemical processes occurring within the soil zone. The meteorologic input of $H^+$ at Hubbard Brook is mainly in the form of $H_2SO_4$ and $HNO_3$ (Tables 10 and 11). The average external net input of hydrogen ion equivalent observed at Hubbard Brook over the past decade is $0.865 \pm .033 \times 10^3$ Eq/ha·yr (Table 11). If this were the only source of $H^+$ ions at Hubbard Brook (and the ecosystem were in a steady state) we might expect this value for hydrogen fixation to be more or less balanced by the net rate at which ionic Ca, Mg, K, Na, and Al were leached from the system. In fact, there are more of these cations removed from the ecosystem each year ($1.267 \pm 0.071 \times 10^3$ Eq/ha) than there are external hydrogen ions to replace them (Table 11). The difference ($0.402 \pm .078 \times 10^3$ Eq/ha·yr) is statistically significant and implies the action of internally generated $H^+$. By this definition "internal" is taken to mean everything that is not sensibly added by meteorologic input.

Some of the internal sources of $H^+$ at Hubbard Brook are as follows.

*Nitrogen.* An interesting aspect of the chemical weathering reaction at Hubbard Brook is the role of the ammonium ion. It is well known to agriculturalists that additions of ammonium ion have an acidifying effect on soils (Brady, 1974; Reuss, 1976). When ammonium ion is fully oxidized to nitrate ion two hydrogen ions are produced. Ammonium is added to the ecosystem both by meteorologic input and by decomposition within the system. The actual acidifying effect of $NH_4^+$ in the Hubbard Brook soils is unknown; however, it is estimated that only about 10 to 20% of the soil ammonium is oxidized (Melillo, 1977). The rest is apparently taken up directly by plants, which may release additional $H^+$ (Reuss, 1976). Although nitrification represents a substantial source of $H^+$, it is still not sufficient by itself to make up the observed deficit between cationic output and external $H^+$ input.

*Carbon.* Reduced carbon in the form of CHO compounds is continually oxidized in the soil zone with the production of $CO_2$ at partial pressures usually exceeding those found in the atmosphere. Some $CO_2$ reacts with the soil water to form carbonic acid, which has a potential to generate $H^+$ ions. However, the soils at Hubbard Brook, at least in the upper horizons, are quite acid (pH $<$ 5). Under these conditions $H_2CO_3$ dissociates only very slightly and hence carbonation weathering reactions (Carroll, 1970) are inhibited. An additional potential source of $H^+$ at Hubbard Brook is from the production of organic acids, such as citric, tartaric, tannic, and oxalic acid, by biologic activity within the soil zone. The role of such organic acids in chemical weathering is undoubtedly important in the Hubbard Brook ecosystem. However, the quantitative effects that these organic acids exert cannot be assessed with our present experimental data.

*Sulfur.* There are small amounts of sulfide minerals in the bedrock at Hubbard Brook that are subject to oxidation in the soil zone with the concommittant production of sulfuric acid (Carroll, 1970). In addition microbially decomposed organic sulfur compounds, such as sulfur, sulfides, thiosulfates and polythionates, are subject to oxidation with the production of sulfuric acid or sulfates.

Oxidation may occur by strict chemical reactions but most sulfur oxidation occurring in the soil profile is thought to be biochemical in nature (Brady, 1974).

**Estimates of weathering rates**

As mentioned above the long-term cationic denudation in the Hubbard Brook ecosystem is measured at $1.267 \pm .071 \times 10^3$ Eq/ha·yr. This situation is somewhat more complicated, however, because we know that both living and dead biomass is accumulating within the undisturbed forest ecosystems at Hubbard Brook. In effect, this net accretion of biomass represents a long-term sink for some of the nutrients supplied from the weathering reaction. The total amount of cations sequestered by this means is $0.722 \times 10^3$ Eq/ha·yr.

This disposition of cations within the ecosystem allows some interesting conclusions. (a) Cations stored within the biomass must be included in calculations of contemporary weathering; (b) the rate of storage is a consequence of the ecosystem's current state of forest succession and so changes with time; and (c) existence of the forest and its state of development must be included in geologic estimates of weathering.

Based on biomass considerations, to be reported more fully in our second volume, we can estimate the role of biomass accumulation in weathering. As indicated previously the HBEF is not presently in a steady state with respect to biomass. It is recovering from a logging operation some 55 yr ago and currently is accumulating biomass, which includes a cationic component. Until the time that steady-state conditions do pertain within the forest some of nutrients provided by chemical weathering must continue to be taken up and stored by the biologic system (Johnson, 1971). If this appraisal of the biologic system at Hubbard Brook is correct, then it follows that the flux of cationic nutrients being diverted into biomass accretion ($0.722 \times 10^3$ Eq/ha·yr) must be added to that actually removed from the system in the form of dissolved load ($1.267 \times 10^3$ Eq/ha·yr) and particulate organic matter ($\sim 0.01 \times 10^3$ Eq/ha·yr). Therefore, our best estimate of cationic denudation (net loss from ecosystem plus long-term storage within the system) at Hubbard Brook becomes about $2.0 \times 10^3$ Eq/ha·yr.

Because the cationic denudation values presented here include biomass accumulation, they are about twice those reported previously (Johnson et al., 1972). For comparison purposes, it is noteworthy that the annual cationic denudation we have determined for Hubbard Brook ($2.0 \times 10^3$ Eq/ha) is comparable to that calculated for New England as a whole ($2.2 \times 10^3$ Eq/ha) but substantially less than the average ($3.8 \times 10^3$ Eq/ha) for the North American continent (Reynolds and Johnson, 1972).

Although it is interesting to note the essential concordance in cationic denudation rates between Hubbard Brook and values for the New England region as a whole, the reason for this concordance is not so clear. The Hubbard Brook data relate to an upland stream system that is sensibly unpolluted and that has a basin composed of igneous and metamorphic rocks. The cationic denudation here is strictly derived from chemical weathering activity. In contrast, the regional values of cationic denudation have been estimated from the large, lowland streams of New England, which flow through long reaches of urban areas and sedimentary rocks. Ostensibly, these streams have abundant opportunities to pick up dissolved loads of road salt, pollutants, and congruently soluble natural salts, all of which do not reflect chemical weathering activity per se. The similarity then between the cationic denudation values for Hubbard Brook and the New England region may be fortuitous, relating as they do to entirely different sets of chemical provenance and pollutive conditions. The similarity also begs an intriguing question: Why is the cationic denudation of lowland basins so low if, in fact, they are highly polluted? Although the final answer here awaits further work, present data suggest that lowland basins should manifest lower cationic denudation compared to highland basins under natural conditions (Reynolds and Johnson, 1972). In highly industrialized regions cationic denudation is probably more an artifact of man's activities than it is nature's. In New England, then, such small upland streams as Hubbard Brook more accurately reflect the natural state of cationic denudation of the landscape than do larger streams.

Our long-term estimates of cationic denudation at Hubbard Brook, then, allow us to estimate the relative importance of external and internal sources of $H^+$ ions as weathering agents. Our evaluation of cationic denudation is $2.0 \times 10^3$ Eq/ha·yr, which should be balanced by the sum of the net external and net internal supply of $H^+$ ions (Figure 29). The external supply rate is $\sim 1.0 \times 10^3$ Eq/ha·yr (Table 11) and by difference the internal source becomes $\sim 1.0 \times 10^3$ Eq/ha·yr. This suggests that under prevailing biologic and chemical conditions external and internal generation of H$^+$ ions play about equal roles in driving the weathering reaction at Hubbard Brook. This conclusion is based on the assumption that all H$^+$ ions entering the ecosystem from both external and internal sources are consumed in the weathering reactions.

**Weathering mode**  Cationic denudation (net export plus biomass accumulation) values accurately portray the quantitative dimensions of chemical weathering but do not disclose any information regarding the qualitative aspects of chemical weathering. For some insight into the latter problem at Hubbard Brook we can look at the relative rates at which various rock-derived elements are weathered. The chemical weathering for elemental Ca, Na, Mg, K, Si, and Al at Hubbard Brook, 21.1, 5.8, 3.5, 7.1, 18.1, and 1.9 kg/ha·yr, respectively, are in serial order Ca > Si > K > Na > Mg > Al (Table 18). This sequence is distinctly different from that predicted by Polynov (1937) for general granitic terrain and that found by Anderson and Hawkes (1958) for New England in particular. It is significant that if the biomass accretion component is neglected, the relative order changes substantially. That is, the net loss of elemental Ca, Na, Mg, K, Si, and Al in dissolved load only is 11.6, 5.6, 2.6, 1.0, 18.0, and 1.9 kg/ha·yr, respectively (Table 11). This results in a serial order of Si > Ca > Na > Mg > Al > K, which is still at considerable variance with

**Table 18.**  Differential Chemical Weathering at Hubbard Brook

| Element | Abundance in bedrock[a], % | A<br>Annual release from bedrock by weathering,[b] kg/ha | B<br>Amount contained in 1,500 kg of bedrock, kg | Differential weathering ratio[c] = (A/B) × 100, % |
|---|---|---|---|---|
| Ca$^{2+}$ | 1.4 | 21.1 | 21.1 | 100 |
| Na$^+$ | 1.6 | 5.8 | 24.1 | 24 |
| Mg$^{2+}$ | 1.1 | 3.5 | 16.5 | 21 |
| K$^+$ | 2.9 | 7.1 | 43.6 | 16 |
| Al$^{3+}$ | 8.3 | 1.9 | 124.8 | 2 |
| Si$^{4+}$ | 30.7 | 18.1[d] | 461.7 | 2 |

[a]Taken from Johnson et al. (1968).

[b]Based on net output of dissolved substances (Table 11) plus living and dead biomass accumulation, which for Ca$^{2+}$, Na$^+$, Mg$^{2+}$, and K$^+$ is 9.5, 0.17, 0.9, and 6.1 kg/ha·yr, respectively.

[c]Normalized to calcium, i.e., assuming the complete extraction of calcium from 1,500 kg of bedrock.

[d]Assuming that dissolved silica is in the form of SiO$_2$ (Table 11), and 0.1 kg Si per hectare is exported in stream water as organic particulate matter.

other published elemental mobility series. These dispari-
ties suggest that differential mobility series have limited
value as a general index to chemical weathering activity,
until such time as collection, sampling, and analytic tech-
niques are normalized to some common basis. The long-
term, comprehensive data collected in the Hubbard Brook
Ecosystem Study allow us to compensate for the effects of
precipitation chemistry, biologic activity, hydrologic dilu-
tion, and seasonal variation in characterizing stream-water
chemistry. To the extent that these factors are crucial in
establishing a relative mobility series, ignoring them must
produce highly unreliable results. Our experience suggests
that the most important factor in this regard is the growth
status of the forest, which may mask the actual rate at
which rocks are chemically decomposing within the
system.

Because of our detailed cationic budget data, however,
we can make other, more direct inferences about differential chemical weathering at Hubbard Brook. The absolute
cationic effluxes we have observed, when compared to the
absolute abundances of these cations in the bedrock at
Hubbard Brook, testify that differential chemical weathering is actively underway (see Table 18). The calcium minerals of the bedrock are obviously the most prone to
weathering. Some 1.5 metric tons per hectare of bedrock
must undergo weathering every year to provide the requi-
site calcium to the system. If we assume that this calcium-
based value (1.5 tons of rock per hectare·year) is the lower
limit for the rate of rock weathering, we can then calculate
the fraction of each element extracted by the weathering
process (Table 18). These latter estimates seem to be
compatible with what is known about the vulnerability of
various silicate minerals to chemical attack (Goldich, 1938;
Marshall, 1964); i.e., calcium-rich minerals are the least
stable and potassium-rich minerals are the most stable in
an acidic chemical regime.

One final point: When compared with the usual published chemical analyses for major streams in New
England (Johnson et al., 1972), the headwater streams of
the HBEF appear to be rather unusual in chemical
makeup; i.e., they are dominated by strong acids (sulfuric
and nitric). Because of its physiographic location, the
HBEF (Figure 2) represents the incipient stages of drainage water; i.e., precipitation has but recently percolated
through the soil zone and is appearing in variable source
areas or first-order stream channels. These acids, how-

ever, are in the process of being neutralized by chemical weathering reactions. The relatively low pH typical for HBEF stream water (pH 4–5) precludes the presence of significant quantities of ionized carbonic acid (Stumm and Morgan, 1970). Significantly, the $SO_4^{2-}$ content of New England stream water is fixed at this stage at 4–6 mg/l and it stays essentially at this concentration during the remainder of its transit to the sea (Johnson et al., 1972; Pearson and Fisher, 1971). As the water moves downstream it is progressively neutralized by weathering reactions. As the pH rises above 5, carbonic acid can dissociate and carbonation weathering reactions may proceed. The rate of neutralization varies somewhat as a function of local streamflow rate and local bedrock chemistry but our experience at Hubbard Brook suggests that the major chemical composition is determined within the first few hundred meters of streamflow. Bicarbonate alkalinity predominates downstream as shown by chemical analyses in the mainstem of Hubbard Brook itself.

# 6 Nutrient Cycles

Chemical elements without a prominent gaseous phase at normal biologic temperatures, such as calcium, magnesium, and potassium, have what is referred to as a sedimentary biogeochemical cycle (Odum, 1959). That is, flux and cycling are affected primarily by hydrologic factors (including dissolution, erosion, and sedimentation), landslides, vulcanism, and biologic agents. The presence of a prominent gaseous component in the biogeochemical cycle (e.g., C, N, and S) greatly complicates measurement and analysis, as discussed previously.

**The calcium cycle**  To illustrate a sedimentary cycle, we shall summarize the average (11 yr) biogeochemical relationships for calcium at Hubbard Brook (Figure 30). Bulk precipitation contributes 2.2 kg/ha of calcium to the ecosystem each year. In addition, some 21.1 kg of Ca per hectare·year are released within the ecosystem by weathering. Because this 55-yr-old forest is accumulating biomass, 5.4 kg of Ca per hectare are stored above ground and 2.7 kg/ha·yr below ground in the annual vegetation growth increment. There is an additional net accumulation of 1.4 kg/ha·yr in the dead biomass of the forest floor. The remainder, 13.9 kg/ha·yr, is lost from the system in drainage water. Most (98.6%) of this loss is in the dissolved form. The annual net loss from the ecosystem (11.7 kg/ha) is only 2% of the total available calcium (510 kg/ha) within the ecosystem and about 19% of the amount circulated annually (uptake = 62.2 kg/ha) by the vegetation.

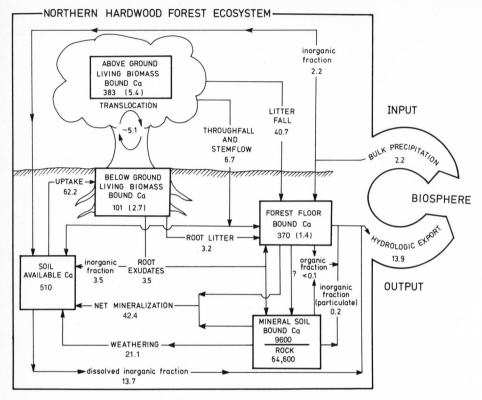

**FIGURE 30.** Annual calcium budget for an aggrading forested ecosystem at Hubbard Brook. Standing crop values are in kg/ha and calcium fluxes are in kg/ha·yr. Values in parentheses represent annual accretion rates.

**FIGURES 30 and 31.** Ecosystem Diagrams.

**General.** The boxes within the diagram represent standing stocks or pools of elements. The values within these boxes are the size of the elemental pool and are in kg/ha. Because the Hubbard Brook Forest is an aggrading ecosystem with an annual net accumulation of biomass, the rate at which the element is accumulating each year is given in parentheses within the appropriate box in kg/ha·yr. The arrows between the various pools represent annual fluxes and are given in kg/ha·yr.

The last cutting of the forest at Hubbard Brook took place between 1909 and 1917. Indications are that the majority of the cutting took place closer to 1917 than to 1909. Therefore, we are making the assumption that forest regrowth was initiated around 1915. The majority of the measurements, which form the basis for the diagrams, were made around 1970. Where the age of the forest was needed (e.g., in the calculation of forest floor accretion) 55 yr was used.

**Pools.** *Aboveground living biomass:* Living biomass was measured by Whittaker et al. (1974). At the same time representative samples were taken and analyzed by Likens and Bormann (1970) for

**FIGURE 31.** Annual sulfur budget for an aggrading forested ecosystem at Hubbard Brook. Standing crop values are in kg/ha and sulfur fluxes are in kg/ha·yr. Values in parentheses represent annual accretion rates.

various elemental constituents. Whittaker et al. (1974) also measured the productivity of the forest for two pentads (1956–1960 and 1961–1965). The biomass accretion values in the diagrams represent a mean of these two pentads and are multiplied by the elemental concentrations of Likens and Bormann (1970).

*Belowground living biomass:* Belowground biomass was measured and elemental composition determined in the same manner as the aboveground living biomass (Whittaker et al., 1974; Likens and Bormann, 1970).

*Forest floor:* The biomass and the elemental composition of the upper organic layers of the soil (to an average depth of 8.9 cm) was measured by Gosz et al. (1976) and Dominski (1971). The biomass of larger litter (e.g., limbs and boles of trees) on the forest floor was measured by Covington (1976) and its elemental composition was determined from Gosz et al. (1972).

The rate of forest floor accretion was estimated by Covington (1976) by measuring the biomass of a successional sequence of northern hardwood stands ranging in age from 3 to 200 yr. Elemental composition is from Gosz et al. (1976).

In some northern hardwood ecosystems, biomass accretion occurs in the form of dead wood which is not part of the living biomass and not yet part of the forest floor. Input–output–standing crop relationships of the dead wood pool at Hubbard Brook are currently under study and preliminary data suggest that dead wood in the 55-yr ecosystem is in steady state where input approximately equals output.

*Mineral soil and rock:* The amount of organic and inorganic matter in the mineral soil and of rock in the soil profile to a depth of 45 cm was measured by Dominski (1971). The elemental composition of mineral soil is from Gosz et al. (1976), Lunt (1932), and Johnson et al. (1968); the elemental composition of rock is from Johnson et al. (1968) assuming 50% unweathered Kinsman quartz monzonite and 50% unweathered Littleton gneiss. Although an arbitrary soil depth of 45 cm was used, we believe the majority of the active soil processes (biogeochemical) occur within the upper 45 cm of soil. During the period of this study we assume that changes in the chemical content of the mineral soil are very small and do not significantly affect the budget calculations.

*Available soil nutrients:* Estimated from soil depth, bulk density, and elemental analyses (Lunt, 1932).

**Elemental Fluxes.** *Bulk precipitation:* Chemical inputs in rain and snow to the ecosystem have been measured since 1963. Numbers used here are weighted mean annual inputs (see Table 10).

*Net gaseous uptake and impaction of aerosols:* This was calculated as net uptake = Losses (net biomass accumulation + net forest floor accumulation + export of dissolved substances and particulate matter) − gains (weathering + precipitation input).

*Litterfall:* The fall of litter from forest vegetation to the forest floor was measured and its elemental composition, determined by Gosz et al. (1972).

*Translocation:* the movement of chemicals back into woody tissues at the time of autumn senescence for leaves, is calculated as translocation = (production + bud litter + root exudates + throughfall and stemflow) − (uptake − shrub and herb litter).

*Throughfall and stemflow:* The washing and leaching of chemicals from the forest canopy by precipitation was determined by Eaton et al. (1973).

*Root exudation:* The loss of chemicals from roots by root exudates was measured by Smith (1976).

*Root litter:* The biomass of root litter was estimated by Whittaker et al. (1974). Elemental composition of roots was taken from Likens and Bormann (1970).

*Net mineralization:* This transformation of organically bound element to inorganic form assumes that there is no microbial immobilization of the element. Net mineralization = (annual uptake by plants + annual hydrologic export of inorganic element) − (annual precipitation input of inorganic element + annual transfer of inorganic element in root exudates + annual transfer of inorganic element in throughfall and stemflow + weathering).

*Uptake:* Uptake by forest vegetation was calculated as uptake = litterfall + net throughfall and stemflow + root exudates + root litter + living biomass accumulation.

*Hydrologic export:* Export of dissolved chemicals represent

means for the past decade (Table 10). Particulate losses represent
means for 4 yr (Table 9).

*Weathering:* Calculated on the basis of 1,504 kg of rock
weathered each year. Elemental composition of the rock substrate is
given by Johnson et al. (1968). See text for further details.


Most of the calcium in the ecosystem, more than 99%,
is present in the soil complex, whereas only about 0.5%
(484 kg/ha) is bound in the vegetation. Nevertheless, about
12% of the available nutrient pool is cycled (taken up) by
the vegetation each year. Meteorologic input accounts for
only about 3.5% of all calcium taken up by vegetation
annually, whereas weathering generates about 34% of
the annual vegetation uptake. Of the annual amount of
calcium taken up by vegetation, litter, stemflow, throughfall, and root exudates return some 87% to the forest floor;
13% of the uptake is accumulated by the forest vegetation.
Interestingly, the living vegetation releases more calcium
in root exudates than is lost as root litter. Gosz et al. (1973)
calculated that some 17.3 kg of Ca per hectare were
released by decomposition and leaching of current litter
each year at Hubbard Brook. Net mineralization from the
entire soil mass is estimated at 42.4 kg of Ca per hectare
(Figure 30). Of the five inputs to the available nutrient
compartment, net mineralization is the largest (56% of the
total), followed by weathering (28%), leaching from the
forest canopy (9%), root exudates (5%), and precipitation
(3%). Although there are large quantities of calcium made
available and circulated within the intrasystem cycle of the
ecosystem, the growing undisturbed forest has minimal
losses. That is, the undisturbed forested ecosystem is very
efficient in cycling and retaining calcium.

Some of the values in Figure 30 differ from those published previously (Likens and Bormann, 1972). This primarily is the result of new or additional data for the living
and dead biomass components of the HBEF and also
because the soil depth used here is only 45 cm instead of 61
cm as used previously.

**The sulfur cycle** Such elements as carbon, sulfur, and nitrogen have a
prominent gaseous phase at normal biologic temperatures,
and this greatly complicates quantitative evaluations of the
biogeochemical flux. To provide an example of this type of
biogeochemical relationship, we shall use sulfur (Figure
31).

The biogeochemistry of sulfur plays a key role in the function of forested ecosystems at Hubbard Brook. Sulfate is the dominant anion in both precipitation and stream water. Approximately 99% of the sulfur in the ecosystem at any one moment is found in the soil complex and the remaining 1% occurs in the living biomass. The vegetation takes up sulfur at a rate of 28.5 kg/ha·yr and accretes it at a rate of 1.2 kg/ha·yr. Weathering generates an amount of sulfur equivalent to about 3% of all sulfur taken up annually by vegetation and meteorologic input accounts approximately 66% of the annual vegetation uptake (Figure 31). The sulfur budget of the forested ecosystems at Hubbard Brook is therefore dominated by meteorologic inputs.

A summation of the values for biomass accretion, weathering, precipitation input, and stream-water output shows a sulfur inbalance of 6.1 kg/ha·yr, suggesting some additional source(s) of sulfur for the ecosystem. We believe that impaction and direct absorption of gaseous sulfur may provide the necessary net sulfur input of 6.1 kg/ha·yr (Figure 31). It is well known that plants can utilize $SO_2$ directly from the atmosphere (Cowling et al., 1973; Hill, 1971; Hoeft et al., 1972), and various authors have suggested input from aerosols and dust impacted on vegetation surfaces (Carlisle et al., 1967; Duvigneaud and-Denaeyer-DeSmet, 1964; Eriksson, 1952; Tamm and Troedsson, 1955; White and Turner, 1970). Our through-fall and stemflow data (Table 6) suggest that sulfurous aerosols may be impacted on vegetation surfaces in large amounts during the summer. We do not know the actual proportions of these two meteorologic sources of sulfur at Hubbard Brook but we estimate that the net gaseous input is about double that from aerosol deposition on an annual basis (Eaton et al., 1976). The combination of net gaseous uptake and aerosol deposition account for about 32% of the total meteorologic input or 35% of the annual hydrologic export. Obviously such inputs must be carefully assessed in biogeochemical studies.

In contrast to its calcium cycle, the ecosystem is not nearly as efficient in retaining sulfur. Some 94% of the sulfur added to the ecosystem annually is lost in stream water.

Quantitative measures of gaseous flux, microbial transformations, and soil standing stocks have not been fully evaluated. These are some of the objectives of ongoing studies at Hubbard Brook.

**Nutrient budget relationships at Hubbard Brook**

The allocation of budgetary items for the major nutrients at Hubbard Brook is given in Tables 19 and 20. These values, along with those of Tables 10–12, suggest several important conclusions regarding the biogeochemistry of the undisturbed northern hardwood ecosystem. These are: (a) In relation to gross losses, nutrient inputs in bulk precipitation represent a significant addition to the ecosystem. Such inputs can be the major source of nutrients in terrestrial ecosystems low in weathering substrates (e.g., Art et al., 1974). (b) For some elements, such as N, S, and Cl, impaction or biologic gaseous fixation can provide significant input to undisturbed ecosystems. (c) Based on these budgets, the undisturbed forest ecosystem shows absolute gains in C, N, S, P, and Cl and absolute losses in Si, Ca, Na, Al, Mg, and K. Losses of the latter substances from the intrasystem nutrient cycle are made up by weathering primary minerals (Figure 1) and the losses therefore imply a decrease in weathering substrate. (d) Weathering is the major source of calcium, potassium, magnesium, and sodium in the ecosystem. (e) assuming that the min-

**Table 19.** Standing Stocks and Annual Biogeochemical Fluxes for a 55-yr old Undisturbed Forested Ecosystem at Hubbard Brook

| Component | Ca | Mg | Na | K | N | S | P | Cl |
|---|---|---|---|---|---|---|---|---|
| | | | | Standing stock (kg/ha) | | | | |
| Aboveground biomass | 383 | 36 | 1.6 | 155 | 351 | 42 | 34 | * |
| Belowground biomass | 101 | 13 | 3.8 | 63 | 181 | 17 | 53 | * |
| Forest floor | 372 | 38 | 3.6 | 66 | 1256 | 124 | 78 | * |
| | | | | Annual flux (kg/ha·yr) | | | | |
| Bulk precipitation input | 2.2 | 0.6 | 1.6 | 0.9 | 6.5 | 12.7 | 0.04 | 6.2 |
| Gaseous or aerosol input | * | * | * | * | 14.2 | 6.1 | * | ? |
| Weathering release | 21.1 | 3.5 | 5.8 | 7.1 | 0 | 0.8 | ? | * |
| Streamwater output | | | | | | | | |
| Dissolved substances | 13.7 | 3.1 | 7.2 | 1.9 | 3.9 | 17.6 | 0.01 | 4.6 |
| Particulate matter | 0.2 | 0.2 | 0.2 | 0.5 | 0.1 | <0.1 | 0.01 | * |
| Vegetation uptake | 62.2 | 9.3 | 34.8 | 64.3 | 79.6[a] | 24.5[a] | 8.9 | * |
| Litter fall | 40.7 | 5.9 | 0.1 | 18.3 | 54.2 | 5.8 | 4.0 | * |
| Root litter | 3.2 | 0.5 | 0.01 | 2.1 | 6.2 | 0.6 | 1.7 | * |
| Throughfall and stemflow | 6.7 | 2.0 | 0.3 | 30.1 | 9.3 | 21.0 | 0.7 | 4.4 |
| Root exudates | 3.5 | 0.2 | 34.2 | 8.0 | 0.9 | 1.9 | 0.2 | 1.8 |
| Net mineralization | 42.4 | 6.1 | 0.1 | 20.1 | 69.6 | 5.7 | ? | ? |
| Aboveground biomass accretion | 5.4 | 0.4 | 0.03 | 4.3 | 4.8 | 0.8 | 0.9 | * |
| Belowground biomass accretion | 2.7 | 0.3 | 0.12 | 1.5 | 4.2 | 0.4 | 1.4 | * |
| Forest floor accretion | 1.4 | 0.2 | 0.02 | 0.3 | 7.7 | 0.8 | 0.5 | * |

*small, unmeasured

[a]root uptake

**Table 20.** Allocation of Budgetary Items in Percent for Watershed Ecosystems of the Hubbard Brook Experimental Forest

|  | Ca | K | Mg | Na | N | S |
|---|---|---|---|---|---|---|
| Source |  |  |  |  |  |  |
| Precipitation input | 9 | 11 | 15 | 22 | 31 | 65 |
| Net gas or aerosol input | — | — | — | — | 69 | 31 |
| Weathering release | 91 | 89 | 85 | 78 | — | 4 |
| Storage or loss |  |  |  |  |  |  |
| Biomass accumulation |  |  |  |  |  |  |
| Vegetation | 35 | 68 | 17 | 2 | 43 | 6 |
| Forest floor | 6 | 4 | 5 | <1 | 37 | 4 |
| Streamflow |  |  |  |  |  |  |
| Dissolved substances | 59 | 22 | 74 | 95 | 19 | 90 |
| Particulate matter | <1 | 6 | 5 | 3 | <1 | <1 |

eral horizons of the soil are near steady state, a large percentage (greater than 40%) of potassium, nitrogen and calcium input from meteorologic sources and released by weathering is stored annually by living and dead biomass within the ecosystem. The forest floor is a particularly effective storage site for nitrogen. In contrast, only about 2% of the sodium and 10% of the sulfur added to the ecosystem is stored in the biomass, and the remainder, more than 90%, is lost in streamflow. (f) Root exudates are very important to the intrasystem cycle of sodium, whereas stemflow and throughfall are particularly important for cycling sulfur and potassium, and litter plays a major role in the cycling of phosphorus, nitrogen, magnesium and calcium.

# 7

## The Northern Hardwood Ecosystem at Hubbard Brook in Relation to Other Forested Ecosystems in the World

Forests and woodlands cover some $57 \times 10^6$ km$^2$, which is about 38% of the total continental area or 11% of the earth's surface. Despite this relatively small area, 92% of the earth's plant biomass and 46% of its annual net primary production come from forests (Table 21). The 79.9 billion metric tons of dry plant matter produced (net) each year by forests exceeds the total net primary production of the oceans, even though the oceans are more than six times larger in area. The animal biomass in these forests represents 68% of the total found on continental land masses (Whittaker and Likens, 1973).

The structure and well-being of forested ecosystems to a large extent reflect the chemical balance that is extant for the ecosystem. It is of interest to examine briefly how the biogeochemical data from the Hubbard Brook Ecosystem Study relate to some other studies done at various places in the world. Budgets of precipitation input/stream-water output for a variety of forested ecosystems of the world are given in Table 22. These data describe the general nutrient flux for these systems.

Data in Table 22 encompass a wide range in ecosystem analysis and sophistication. In some cases these data represent averages of several watersheds for several years. In others, data are scanty and are derived from a poorly defined hydrologic system. These data, therefore, although very interesting and informative, should be treated with some caution. Moreover, these relationships may become modified as more information on net

**Table 21.**  Estimates of Area, Biomass, and Productivity for Major Components of the Biosphere[a]

| | Area, $10^6$ km$^2$ | Plant biomass, $10^9$ dry tons | Net plant productivity, $10^9$ dry tons/yr | Animal biomass, $10^6$ dry tons | Animal productivity, $10^6$ dry tons/yr |
|---|---|---|---|---|---|
| Forest and woodlands | 57 | 1,700 | 79.9 | 680 | 442 |
| Total continental area | 149 | 1,837 | 117.5 | 1,015 | 827 |
| Total ocean | 361 | 4 | 55.0 | 998 | 3,057 |
| Total biosphere | 510 | 1,841 | 172.5 | 2,013 | 3,884 |

[a]It is assumed that dry matter $\times$ 0.45 = carbon. Modified from Whittaker and Likens (1973, 1975).

exchange of gases and dry deposition of aerosols becomes available. We have not incorporated adjustments for net gaseous input and dry deposition of aerosols in Table 22, except in the case of N and S for Hubbard Brook, for which data have been obtained from an analysis of the total ecosystem. Considerable effort should be devoted to the problem of measuring dry deposition and gaseous exchange in forested ecosystems in the future.

In general, the findings from these studies show a common pattern with those of the Hubbard Brook Ecosystem Study. Several conclusions emerge: (1) Chemicals in precipitation are significant additions to the budgets for a variety of forested ecosystems. (2) Precipitation inputs of inorganic nitrogen and phosphorus exceed losses in stream water for undisturbed, forested ecosystems. There are three exceptions to the generalization that phosphorus inputs in precipitation exceed outputs. Although most of the watersheds in the Western Cascades Range of Oregon showed outputs larger than inputs during the years studied, some watersheds showed net gains (Fredriksen, 1975) and the pattern was therefore not clear. The other exceptions, Finland and Taughannock Creek, NY, include drainage from agricultural land. Man's activities or production of wastes can readily overwhelm the capacity of natural ecosystems to conserve phosphorus within the ecosystem (Likens and Bormann, 1974a). (3) In general, there are net losses of calcium, magnesium, sodium, and potassium from forested ecosystems that must be made up by weathering. (4) Geologic substrates play a major role in determining the amount and composition of chemicals lost in stream water. Watersheds on relatively soluble bedrock (England; Taughannock Creek, NY; Walker Branch, TE;

**Table 22.** Nutrient Budgets for Various Terrestrial Ecosystems of the World (kg/ha·yr)

| Location[a] | Precipitation input | Stream-water output | Net gain or loss | Annual precipitation, cm | Dominant vegetation[b] | Geology[c] |
|---|---|---|---|---|---|---|
| **Calcium** | | | | | | |
| **Temperate: mostly angiosperm and deciduous forest** | | | | | | |
| Coweeta, NC, U.S. (1) | 6.2 | 6.9 | −0.7 | 185 | Q, C, Ac | Mg, Msh |
| Hubbard Brook, U.S. (2) | 2.2 | 13.9 | −11.7 | 130 | Ac, F, B | Ig, Msh |
| Long Island, NY, U.S. (3) | 3.3 | 9.6d | −6.3 | 124 | Q, Pin | Ss |
| Pago Catchment, Australia (4) | 7.0 | 7.1 | −0.1 | 150 | E | Msh, Ig |
| S.E. U.S. (5) | 6 | 19 | −13 | 127 | Q, Pin | Ss, Ssh, Msh |
| Silverstream, New Zealand (6) | 6 | 26 | −20 | 135 | N | Ss, Ssh, Msh, Msh |
| Taughannock Creek, NY, U.S. (7) | 14.9 | 182 | −167 | 96 | Ac, Ti, Ts | Ssh, Sc |
| Tesuque Watersheds, NM, U.S. (8) | 6.8 | 13.1 | −6.3 | 72 | Po | Ig, Mg |
| Walker Branch, TE, U.S. (9) | 14.3 | 148 | −134 | 155 | Q, C | Sc |
| **Temperate: mostly coniferous and evergreen forest** | | | | | | |
| Birkenes Watershed, Norway (10) | 2.7 | 14.1 | −11.4 | 134 | B, Pin, Pic | Ig |
| Blue Range Catchment, Australia (4) | 6.2 | 5.5 | +0.7 | 132 | Pin | Ib, Msh |
| Carnation Creek, Vancouver Island, Canada (11) | 3.7 | 57.7 | −54 | 315 | Ab, Ts, Th | Mb, Ms, Mc |
| Cedar River, WA, U.S. (12) | 2.8 | 4.5d | −1.7 | 136 | Ps | Ss |
| ELA, Ontario, Canada (13) | 3.8 | 6.0 | −2.2 | 83 | Pin, Pic | Ig, Mg |
| Finland (14) | 2 | 12 | −10 | 57 | Pin, Pic, B | Ig, Mg |
| Storsjön, Sweden (15) | 3.6 | 12.2 | −8.6 | 87 | Pin, Pic | Ig |
| Tesuque Watersheds, NM, U.S. (8) | 6.4 | 10.1 | −3.7 | 74 | Pic, Ab, Pin | Ig, Mg |
| Tesuque Watersheds, NM, U.S. (8) | 6.7 | 22.3 | −15.6 | 48 | Pin, J | Ig, Mg |
| Velen, Sweden (16) | 5.9e | 11.1e | −5.2 | 72 | Pin, Pic | Mg, Ig |
| **Temperate: bog vegetation** | | | | | | |
| Rough Sike Catchment, England (17) | 9.0 | 53.8f | −44.8 | 213 | Bog | Sc, Ss, Ssh |
| **Tropical: angiosperm, mostly evergreen forest** | | | | | | |
| El Verde, Puerto Rico (18) | 21.8 | 43.1 | −21.3 | 376 | Trp | Ib, Ssh, Ss |
| Rio Negro, Brazil (19) | 0.8 | 4.7g | −3.9 | 200 | Trp | Ssh, Msh, Mg |

**Table 22.**  Nutrient Budgets for Various Terrestrial Ecosystems of the World (kg/ha·yr) (continued)

| Location[a] | Precipitation input | Stream-water output | Net gain or loss | Annual precipitation, cm | Dominant vegetation[b] | Geology[c] |
|---|---|---|---|---|---|---|
| **Magnesium** | | | | | | |
| **Temperate:** mostly angiosperm and deciduous forest | | | | | | |
| Coweeta, NC, U.S. | 1.3 | 3.1 | −1.8 | 185 | Q, C, Ac | $M_g$, $M_{sh}$ |
| Hubbard Brook, U.S. | 0.6 | 3.3 | −2.7 | 130 | Ac, F, B | $I_g$, $M_{sh}$ |
| Long Island, NY, U.S. | 2.1 | 7.3[d] | −5.2 | 124 | Q, Pin | $S_s$ |
| Pago Catchment, Australia | 1.5 | 4.5 | −3.0 | 150 | E | $M_{sh}$, $I_g$ |
| S.E. U.S. | 2 | 6 | −4 | 127 | Q, Pin | $S_s$, $S_{sh}$, $M_{sh}$ |
| Silverstream, New Zealand | 9 | 13 | −4 | 135 | N | $S_s$, $S_{sh}$, $M_s$, $M_{sh}$ |
| Taughannock Creek, NY, U.S. | 1.4 | 34.8 | −33.4 | 96 | Ac, Ti, Ts | $S_{sh}$, $S_c$ |
| Tesuque Watersheds, NM, U.S. | 0.7 | 4.2 | −3.5 | 72 | Po | $I_g$, $M_g$ |
| Walker Branch, TE, U.S. | 2.1 | 77.1 | −75.0 | 155 | Q, C | $S_c$ |
| **Temperate:** mostly coniferous and evergreen forest | | | | | | |
| Birkenes Watershed, Norway | 2.3 | 5.0 | −2.7 | 134 | B, Pin, Pic | $I_g$ |
| Blue Range Catchment, Australia | 1.3 | 8.7 | −7.4 | 132 | Pin | $I_g$, $M_{sh}$ |
| Carnation Creek, Vancouver Island | 3.0 | 10.4 | −7.4 | 315 | Ab, Ts, Th | $M_b$, $M_{si}$, $M_c$ |
| ELA, Ontario, Canada | 0.9 | 2.4 | −1.5 | 83 | Pin, Pic | $I_g$, $M_g$ |
| Finland | 1 | 4 | −3 | 57 | Pin, Pic, B | $I_g$, $M_g$ |
| Storsjön, Sweden | 1.9 | 8.7 | −6.8 | 87 | Pin, Pic | $I_g$ |
| Tesuque Watersheds, NM. U.S. | 0.6 | 3.4 | −2.8 | 74 | Pic, Ab, Pin | $I_g$, $M_g$ |
| Tesuque Watersheds, NM. U.S. | 0.7 | 10.8 | −10.1 | 48 | Pin, J | $I_g$, $M_g$ |
| Velen, Sweden | 1.2[e] | 3.2[e] | −2.0 | 72 | Pin, Pic | $M_g$, $I_g$ |
| **Temperate:** bog vegetation | | | | | | |
| Maesnant Catchment, mid-Wales[a] (20) | 4.4 | 8.7 | −4.3 | 164 | Bog | $S_{sh}$, $S_s$ |
| **Tropical:** angiosperm, mostly evergreen forest | | | | | | |
| El Verde, Puerto Rico | 4.9 | 15.0 | −10.1 | 376 | Trp | $I_b$, $S_{sh}$, $S_s$ |
| Rio Negro, Brazil | 2.0 | 3.0[g] | −1.0 | 200 | Trp | $S_{sh}$, $M_{sh}$, $M_g$ |

**Sodium**

| | | | | | | |
|---|---|---|---|---|---|---|
| **Temperate: mostly angiosperm and deciduous forest** | | | | | | |
| Coweeta, NC, U.S. | 5.4 | 9.7 | -4.3 | 185 | Q, C, Ac | $M_g$, $M_{sh}$ |
| Hubbard Brook, U.S. | 1.6 | 7.5 | -5.9 | 130 | Ac, F, B | $I_g$, $M_{sh}$ |
| Long Island, NY, U.S. | 17 | 23$^d$ | -6 | 124 | Q, Pin | $S_s$ |
| Pago Catchment, Australia | 2.5 | 15.9 | -13.4 | 150 | E | $M_{sh}$, $I_g$ |
| S.E. U.S. | 5 | 26 | -21 | 127 | Q, Pin | $S_s$, $S_{sh}$, $M_{sh}$ |
| Silverstream, New Zealand | 50 | 62 | -12 | 135 | N | $S_a$, $S_{sh}$, $M_s$, $M_{sh}$ |
| Taughannock Creek, NY, U.S. | 2.3 | 18.9 | -16.6 | 96 | Ac, Ti, Ts | $S_{sh}$, $S_c$ |
| Tesuque Watersheds, NM, U.S. | 0.8 | 10.0 | -9.2 | 72 | Po | $I_g$, $M_g$ |
| Walker Branch, TE, U.S. | 3.9 | 4.5 | -0.6 | 155 | Q, C | $S_c$ |
| **Temperate: mostly coniferous and evergreen forest** | | | | | | |
| Birkenes Watershed, Norway | 19.1$^h$ | 30.3 | -11.2 | 134 | B, Pin, Pic | $I_g$ |
| Blue Range Catchment, Australia | 2.2 | 16.2 | -14.0 | 132 | Pin | $I_b$, $M_{sh}$ |
| Carnation Creek, Vancouver Island, Canada | 11.0 | 38.4 | -27.4 | 315 | Ab, Ts, Th | $M_b$, $M_s$, $M_c$ |
| ELA, Ontario, Canada | 1.6 | 3.7 | -2.1 | 83 | Pin, Pic | $I_g$, $M_g$ |
| Finland | 2 | 6 | -4 | 57 | Pin, Pic, B | $I_g$, $M_g$ |
| Storsjön, Sweden | 12.9 | 37.5 | -24.6 | 87 | Pin, Pic | $I_g$ |
| Tesuque Watersheds, NM, U.S. | 1.0 | 6.5 | -5.5 | 74 | Pic, Ab, Pin | $I_g$, $M_g$ |
| Tesuque Watersheds, NM, U.S. | 0.7 | 4.9 | -4.2 | 48 | Pin, J | $I_g$, $M_g$ |
| Velen, Sweden | 3.9$^e$ | 9.2$^e$ | -5.3 | 72 | Pin, Pic | $M_g$, $I_g$ |
| **Temperate: bog vegetation** | | | | | | |
| Maesnant Catchment, mid-Wales | 27.2 | 43.7 | -16.5 | 164 | Bog | $S_{sh}$, $S_s$ |
| Rough Sike Catchment, England | 25.5 | 45.2$^f$ | -19.7 | 213 | Bog | $S_c$, $S_s$, $S_{sh}$ |
| **Tropical: angiosperm, mostly evergreen forest** | | | | | | |
| El Verde, Puerto Rico | 57.2 | 64.5 | -7.3 | 376 | Trp | $I_b$, $S_{sh}$, $S_s$ |

# Table 22. Nutrient Budgets for Various Terrestrial Ecosystems of the World (kg/ha·yr) (continued)

| Location[a] | Precipitation input | Stream-water output | Net gain or loss | Annual precipitation, cm | Dominant vegetation[b] | Geology[c] |
|---|---|---|---|---|---|---|
| **Potassium** | | | | | | |
| **Temperate: mostly angiosperm and deciduous forest** | | | | | | |
| Coweeta, NC, U.S. | 3.2 | 5.2 | −2.0 | 185 | Q, C, Ac | $M_g$, $M_{sh}$ |
| Hubbard Brook, U.S. | 0.9 | 2.4 | −1.5 | 130 | Ac, F, B | $I_g$, $M_{sh}$ |
| Long Island, NY, U.S. | 2.4 | 3.9[d] | −1.5 | 124 | Q, Pin | $S_s$ |
| S.E. U.S. | 1 | 6 | −5 | 127 | Q, Pin | $S_s$, $S_{sh}$, $M_{sh}$ |
| Silverstream, New Zealand | 5 | 13 | −8 | 135 | N | $S_s$, $S_{sh}$, $M_s$, $M_{sh}$ |
| Taughannock Creek, NY, U.S. | 0.8 | 5.6 | −4.8 | 96 | Ac, Ti, Ts | $S_{sh}$, $S_c$ |
| Tesuque Watersheds, NM, U.S. | 1.3 | 3.2 | −1.9 | 72 | Po | $I_g$, $M_g$ |
| Walker Branch, TE, U.S. | 3.1 | 6.8 | −3.7 | 155 | Q, C | $S_c$ |
| **Temperate: mostly coniferous and evergreen forest** | | | | | | |
| Birkenes Watershed, Norway | 2.2 | 1.5 | +0.7 | 134 | B, Pin, Pic | $I_g$ |
| Carnation Creek, Vancouver Island, Canada | 2.5 | 4.8 | −2.3 | 315 | Ab, Ts, Th | $M_b$, $M_s$, $M_c$ |
| Cedar River, WA, U.S. | 0.8 | 1.0[d] | −0.2 | 136 | Ps | $S_s$ |
| ELA, Ontario, Canada | 1.1 | 1.2 | −0.1 | 83 | Pin, Pic | $I_g$, $M_g$ |
| Finland | 2.4 | 4.6 | −2[*] | 57 | Pin, Pic, B | $I_g$, $M_g$ |
| Storsjön, Sweden | 1.7 | 4.0 | −2.3 | 87 | Pin, Pic | $I_g$ |
| Tesuque watersheds, NM, U.S. | 1.4 | 2.1 | −0.7 | 74 | Pic, Ab, Pin | $I_g$, $M_g$ |
| Tesuque watersheds, NM, U.S. | 1.1 | 1.2 | −0.1 | 48 | Pin, J | $I_g$, $M_g$ |
| Velen, Sweden | 1.6[e] | 2.5[e] | −0.9 | 72 | Pin, Pic | $M_{gr}$, $I_g$ |
| **Temperate: bog vegetation** | | | | | | |
| Maesnant Catchment, mid-Wales | 1.6 | 2.6 | −1.0 | 164 | Bog | $S_{sh}$, $S_s$ |
| Rough Sike Catchment, England | 3.1 | 9.0[f] | −5.9 | 213 | Bog | $S_{cr}$, $S_s$, $S_{sh}$ |
| **Tropical: angiosperm, mostly evergreen forest** | | | | | | |
| El Verde, Puerto Rico | 18.2 | 20.8 | −2.6 | 376 | Trp | $I_b$, $S_{sh}$, $S_s$ |

Nitrogen

| | | | | | | |
|---|---|---|---|---|---|---|
| **Temperate:** mostly angiosperm and deciduous forest | Coshocton, OH, U.S. (21) | 20 | 2.5 | +17.5 | 89 | L, Pin | $S_{sh}$, $S_c$, $S_s$ |
| | Hubbard Brook, U.S. | $20.7^{i,j}$ | $4.0^i$ | +16.7 | 130 | Ac, F, B | $I_g$, $M_{sh}$ |
| | S.E. U.S. | $2^k$ | $1^k$ | $+1^k$ | 127 | Q, Pin | $S_s$, $S_{sh}$, $M_{sh}$ |
| | Silverstream, New Zealand | 2.2 | 1.8 | +0.4 | 135 | N | $S_s$, $S_{sh}$, $M_s$, $M_{sh}$ |
| | Taughannock Creek, NY, U.S. | $9.7^i$ | $5.6^i$ | +4.1 | 96 | Ac, Ti, Ts | $S_{sh}$, $S_c$ |
| | Walker Branch, TE, U.S. | 8.7 | 1.8 | +6.9 | 155 | Q, C | $S_c$ |
| **Temperate:** mostly coniferous and evergreen forest | Birkenes Watershed, Norway | $14.5^i$ | $2.2^i$ | +12.3 | 134 | B, 'Pin, Pic | $I_g$ |
| | Carnation Creek, Vancouver Island, Canada | $2.7^i$ | $1.1^i$ | +1.6 | 315 | Ab, Ts, Th | $M_b$, $M_s$, $M_c$ |
| | Cedar River, WA, U.S. | 1.1 | $0.6^d$ | +0.5 | 136 | Ps | $S_s$ |
| | ELA, Ontario, Canada | 6.4 | 0.9 | +5.5 | 83 | Pin, Pic | $I_g$, $M_g$ |
| | Finland | 6 | 2 | +4 | 57 | Pin, Pic, B | $I_g$, $M_g$ |
| | Storsjön, Sweden | 10.0 | 2.3 | +7.7 | 87 | Pin, Pic | $I_g$ |
| | Velen, Sweden | $5.9^{e,i}$ | $0.4^{e,i}$ | +5.5 | 72 | Pin, Pic | $M_g$, $I_g$ |
| | Western Cascades Range, OR, U.S. (22) | $2.5^i$ | $1.2^i$ | +1.3 | 200 | Ps | $I_b$ |
| **Temperate:** bog vegetation | Rough Sike Catchment, England | 8.2 | $3.0^f$ | +5.2 | 213 | Bog | $S_c$, $S_s$, $S_{sh}$ |
| **Tropical:** angiosperm mostly evergreen forest | Rio Negro, Brazil | 5.6 | $4.7^g$ | +0.9 | 200 | Trp | $S_{sh}$, $M_{sh}$, $M_g$ |
| **Tundra:** low shrub vegetation | Kuokkel area, Sweden (23) | $1.15^m$ | $0.97^m$ | +0.18 | 75 | Ca, Em, B | $M_{sh}$ |

# Table 22. Nutrient Budgets for Various Terrestrial Ecosystems of the World (kg/ha·yr) (continued)

| Location[a] | Precipitation input | Stream-water output | Net gain or loss | Annual precipitation, cm | Dominant vegetation[b] | Geology[c] |
|---|---|---|---|---|---|---|
| **Phosphorus** | | | | | | |
| **Temperate: mostly angiosperm and deciduous forest** | | | | | | |
| Coshocton, OH, U.S. | 0.18[n] | 0.05 | +0.13 | 89 | L, Pin | $S_{sh}$, $S_{cr}$, $S_s$ |
| Hubbard Brook, U.S. | 0.036 | 0.019 | +0.017 | 130 | Ac, F, B | $I_g$, $M_{sh}$ |
| Pago Catchment, Australia | 0.33 | 0.26 | +0.07 | 150 | E | $M_{sh}$, $I_g$ |
| Silverstream, New Zealand | 0.2 | 0.03 | +0.2 | 135 | N | $S_s$, $S_{sh}$, $M_{sl}$, $M_{sh}$ |
| Taughannock Creek, NY, U.S. | 0.07 | 0.20 | −0.13 | 96 | Ac, Tl, Ts | $S_{sh}$, $S_c$ |
| Walker Branch, TE, U.S. | 0.54 | 0.02 | +0.52 | 155 | Q, C | $S_c$ |
| **Temperate: mostly coniferous and evergreen forest** | | | | | | |
| Blue Range Catchment, Australia | 0.39 | 0.42 | −0.03 | 132 | Pin | $I_b$, $M_{sh}$ |
| Boundary Waters Canoe Area, MIN, U.S. (24) | 0.14 | 0.015 | +0.13 | 70 | Pin, Pic, B | $I_g$, $S_{sh}$ |
| Carnation Creek, Vancouver Island, Canada | 0.11 | 0.05 | +0.06 | 315 | Ab, Ts, Th | $M_b$, $M_s$, $M_c$ |
| Cedar River, WA, U.S. | o | 0.02 | +? | 136 | Ps | $S_s$ |
| Clear Lake, Ontario, Canada (25) | 0.35 | 0.09 | +0.26 | 90 | Ac, F, Q | $I_g$, $M_g$ |
| ELA, Ontario, Canada | 0.32 | 0.05 | +0.27 | 83 | Pin, Pic | $I_g$, $M_g$ |
| Finland | 0.1 | 0.3 | −0.2 | 57 | Pin, Pic, B | $I_g$, $M_g$ |
| Storsjön, Sweden | 0.14 | 0.02 | +0.12 | 87 | Pin, Pic | $I_g$ |
| Western Cascades Range, OR, U.S. | 0.29[p] | 0.51[p] | −0.22 | 200 | Ps | $I_b$ |
| **Temperate: bog vegetation** | | | | | | |
| Rough Sike Catchment, England | 0.6 | 0.4[f] | +0.2 | 213 | Bog | $S_{cr}$, $S_s$, $S_{sh}$ |
| **Tropical: angiosperm, mostly evergreen forest** | | | | | | |
| Rio Negro, Brazil | 0.2 | 0.1[g] | +0.1 | 213 | Trp | $S_{sh}$, $M_{sh}$, $M_g$ |
| **Tundra: low shrub vegetation** | | | | | | |
| Kuokkel area, Sweden | 0.055[m] | 0.041[m] | +0.014 | 75 | Ca, Em, B | $M_{sh}$ |

## Sulfate–sulfur

**Temperate: mostly angiosperm and deciduous forest**

| Location | | | | | | |
|---|---|---|---|---|---|---|
| Hubbard Brook, U.S. | 18.8[i] | 17.6 | +1.2 | 130 | Ac, F, B | $I_g$, $M_{sh}$ |
| S.E. U.S. | 8 | 7 | +1 | 127 | Q, Pin | $S_s$, $S_{sh}$, $M_{sh}$ |
| Silverstream, New Zealand | 7 | 13 | -6 | 135 | N | $S_s$, $S_{sh}$, $M_s$, $M_{sh}$ |
| Taughannock Creek, NY, U.S. | 21 | 38 | -17 | 96 | Ac, Ti, Ts | $S_{sh}$, $S_c$ |
| Walker Branch, TE, U.S. | 18.8 | 11.3 | +7.5 | 155 | Q, C | $S_c$ |

**Temperate: mostly coniferous and evergreen forest**

| Location | | | | | | |
|---|---|---|---|---|---|---|
| Birkenes Watershed, Norway | 15.6 | 26.9 | -11.3 | 134 | B, Pin, Pic | $I_g$ |
| Carnation Creek, Vancouver Island, Canada | 8.7 | 28 | -19 | 315 | Ab, Ts, Th | $M_b$, $M_s$, $M_c$ |
| ELA, Ontario, Canada | 3.0 | 3.2 | -0.2 | 83 | Pin, Pic | $I_g$, $M_g$ |
| Finland | 1.4 | 4.7 | -3.3 | 57 | Pin, Pic, B | $I_g$, $M_g$ |
| Storsjön, Sweden | 11.5 | 25.3 | -13.8 | 87 | Pin, Pic | $I_g$ |
| Velen, Sweden | 10.3[e] | 9.4[e] | +0.9 | 72 | Pin, Pic | $M_g$, $I_g$ |

[a] (1) National Academy of Sciences (1974). (2) Present study. (3) Woodwell and Whittaker (1967). (4) P. M. Hallam, personal communication (1976). (5) Gambell and Fisher (1966). (6) Schindler Miller (1963). (7) Likens (1972). (8) Gosz (1977). (9) G. Henderson, personal communication (1976). (10) Giessing et al. (1976). (11) Scrivner (1975). (12) Cole et al. (1967). (13) Schindler et al. (1976). (14) Viro (1953). (15) W. Dickson, personal communication (1976). (16) Eriksson (1974). (17) Crisp (1966). (18) Jordan and Kline (1972). (19) Ungemach (1967, 1970). (20) Cryer (1976). (21) Taylor et al. (1971). (22) Fredriksen (1975). (23) M. Jansson, personal communication (1976). (24) Wright (1974). (25) Schindler and Nighswander (1970).

[b] Vegetation footnotes: Ab, *Abies*; Ac, *Acer*; B, *Betula*; Bog, bog species; C, *Carya*; Ca, *Calluna*; E, *Eucalyptus*; Em, *Empetrum*; F, *Fagus*; J, *Juniperus*; L, locust; N, *Nothofagus*; Pin, *Pinus*; Pic, *Picea*; Ps, *Pseudotsuga*; Po, *Populus*; Q, *Quercus*; Th, *Thuja*; Ti, *Tilia*; Ts, *Tsuga*; Trp, tropical species.

[c] Geology footnotes: $I_g$, igneous, granitic; $I_b$, igneous, basaltic; $S_c$, sedimentary, carbonate; $S_{sh}$, sedimentary, shale; $S_s$, sedimentary, sandstone; $M_x$, metamorphic (x refers to subscripts used above).

[d] To water table.

[e] Average for 6 yr, 1968–1973.

[f] Transport of eroded peat not included.

[g] Assuming measurements of discharge and concentration are representative of the entire year.

[h] Calculated from Mg inputs and Na:Mg ratio in seawater.

[i] Also includes net gaseous uptake and aerosol deposition; see Table 10 for bulk precipitation values.

[j] $NO_3$-N + $NH_4$-N.

[k] $NO_3$-N only.

[l] Average of five drainage basins, 1971–1973.

[m] Average for 5 yr, 1971–1975.

[n] Calculated from 89 cm of precipitation times concentration (0.02 mg P/l; Taylor et al., 1971).

[o] "Trace."

[p] Average of five drainage basins, 1972–1973.

[q] Data for calcium omitted because of apparent contamination of precipitation samples.

El Verde, Puerto Rico; and Carnation Creek, Vancouver Island) lose large quantities of the major cations via streamflow. (5) Forested ecosystems in coastal locations or otherwise dominated by maritime air masses have relatively large inputs of sea salts in precipitation. (6) A close correlation between the amount of precipitation and the loss of nutrients may apply to individual watersheds, such as the HBEF, but as a general principle there is no close correlation between the amount of precipitation and loss of nutrients in forested ecosystems with different climates. (7) Net loss or gain values for sulfur budgets are significantly affected by anthropogenic contributions to precipitation. Contaminated air masses over Hubbard Brook; Taughannock Creek, NY; Walker Branch, TE; Birkenes, Norway; and, to a lesser extent, southern Sweden, the southeastern United States, and Vancouver Island, Canada, produce high levels of sulfur in local precipitation. (8) Undisturbed, forested ecosystems in a variety of biomes have conservative losses of nutrients relative to inputs or amounts cycled internally, and (9) Stream-water losses may be very significantly changed by disturbance of the ecosystem, such as forest cutting or other land management practices (see also Likens and Bormann, 1974a).

# 8 Summary Discussion and Conclusions

This book summarizes our current understanding of the biogeochemistry of an undisturbed northern hardwood forest ecosystem at Hubbard Brook. It emphasizes the usefulness of the small watershed technique of biogeochemical measurement in shedding light on ecosystem function.

Trees are long lived and over time accumulate large amounts of biomass as wood; it is therefore not surprising that trees dominate the forest landscape and are used to name the ecosystem types (plant associations or biomes). Man has long looked to the forests for firewood and timber, as well as for clean water, game, and recreation, but his demands often have been at variance with the structure of the forest ecosystem and large areas have been (and are being) cleared of forest vegetation. This is currently proceeding at an unprecedented rate in the tropical forest ecosystems (e.g., Croat, 1972; Gómez-Pompa et al., 1972).

Water in some form is also a conspicuous feature of the humid forested landscape. Rain and snow not only supply the water that replenishes ground water reserves and fills stream and river channels, ponds, lakes, swamps, and marshes but also provide the moisture and many of the nutrients for forest growth. Chemicals, particularly those without a prominent gaseous phase, such as calcium, magnesium, or potassium, are transported largely into and out of ecosystems by moving water.

In forested landscapes much of the water that falls to the land's surface as precipitation is first intercepted by the leaves and branches of the trees and herbaceous layer and then by litter on the ground. Therefore, the kinetic energy and chemical composition of the water may be appreciably altered before the water comes into contact with the mineral soil. This both minimizes erosion and regulates nutrient flow. This ostensibly simple and obvious interaction emphasizes that the regulation of ecosystem processes, such as biogeochemical flux, depends on the integrity of the whole ecosystem.

Water that has fallen to the surface may run off overland (minimal in most forests), infiltrate into the soil, or evaporate. Again, the biotic portion of the forest ecosystem may alter both the distribution of potential and kinetic energy and the chemistry of water circulating within an ecosystem.

There are several ways, then, in which water may affect the biogeochemistry of a forested ecosystem. We have found it useful for quantitative studies to select ecosystems that (1) are watersheds, (2) have a watertight bedrock or other substrate, and (3) have a uniform biogeochemical environment (Bormann and Likens, 1967; Likens and Bormann, 1972).

The flux of water and nutrients across an ecosystem's boundaries, as well as internal cycling, is vital for the maintenance of a natural ecosystem. Environmental stress, such as pollution, deforestation, or road building, will alter these conditions of nutrient flux and cycling and thereby disrupt the function of the undisturbed ecosystem. Too little is known about the long-term effects of such alterations or about the ability of an ecosystem to respond, recover, or improvise following disturbance.

The yield of liquid water in streamflow is an important consideration to watershed managers. For example, Douglass and Swank (1972) state that "Because the difference between precipitation input and vapor loss represents the quantity of water available for man's use, the watershed manager seeks to reduce the total vapor loss from the forest vegetation in order to increase the flow of streams." (p. 1). However, because liquid water that flows out of an ecosystem contains nutrients (whether at a constant concentration or not) it may be argued that transpiration is a nutrient conservation mechanism for the ecosystem. Moreover, as precipitation comes in contact with the vegetation its chemistry is altered significantly. The kinetic energy of the falling raindrop is largely absorbed by the vegetation canopy during the growing season and the potential for erosion is reduced. Litter and roots in the forest soil also reduce erosion. The change of liquid water to vapor by the ecosystem also may have great impact in reducing the potential for erosion from liquid water. Therefore, the biotic structure of these forested ecosystems may alter both the potential energy and the chemistry of water as it passes through the ecosystem. Inadequate consideration has been given to the interplay of such changes in the hydrologic cycle on nutrient flux for forested ecosystems. This is of particular concern when the basic structure of the forest ecosystem is altered, e.g., by vegetation removal (Likens et al., 1970).

There is currently considerable effort to quantify biogeochemical relationships for various kinds of terrestrial ecosystems. As a result of our study, we emphasize that:

(1) Considerable thought be given (i) to selection of the ecosystem, including boundaries, for study—millions of dollars and thousands of hours of time may be wasted by a poor initial choice; and (ii) to methodology, such as frequency and location of sampling; criteria for acceptance or rejection of particular samples; and types and reliability of chemical analyses. This is important because small analytical mistakes can easily be blown out of proportion in the multiplication procedures that are inevitable in the construction of ecosystem budgets.

(2) Ecosystems are open. Water and nutrients continually flux through the boundaries and cycle internally between the various components of the ecosystem. Studies at Hubbard Brook have demonstrated the importance of the abiotic factors and quantitative mass balance studies to understanding the structure and function of ecosystems.

(3) Long-term records are important in identifying patterns of nutrient flux through ecosystems. Flux is greatly influenced by hydrology and, during 11 yr precipitation has ranged from 95 to 186 cm/yr at the HBEF. Measurements in either of these extreme years would have given poor estimates of average conditions. Over a 10-yr period, moreover, annual nitrate and hydrogen ion inputs in precipitation increased 2.3 and 1.4 times, respectively. The output of $NO_3$ in stream water also increased during this period and seemed to be strongly influenced by soil freezing during two of the 11 winters.

(4) Although the chemistry of precipitation is highly variable and dependent on complex meteorologic conditions, meteorologic input is an important source of

chemicals for the ecosystem. For S, N, Mg, Na, Ca, and K, meteorologic input supplies about 66, 26, 6, 5, 4, and 1%, respectively, of the annual uptake by green plant biomass.

(5) Meteorologic input may be strongly influenced by advertent and inadvertent manipulation by man. Acid precipitation is a good example: Most acidity in the rain and snow at Hubbard Brook is traceable to sulfur pollution but within the last 10 yr increases in the input of acid precipitation seem to be primarily related to increases in nitrate resulting from internal combustion engines.

(6) Currently the ecosystem is acting as a filter for atmospheric contaminants, such as $H^+$, N, S, P, and certain heavy metals, which accumulate within the ecosystem. The undisturbed forest ecosystem, therefore, is currently acting as a pollution buffer for society; the ultimate cost of this activity is unknown but it seems to trend toward long-term ecosystem degradation.

(7) Annual evapotranspiration (mostly transpiration from vegetation) at Hubbard Brook is essentially constant over a wide range of precipitation and environmental conditions. Water converted to vapor by evapotranspiration cannot serve to erode the system or transport dissolved substances out of the ecosystem. Evapotranspiration not only serves as a regulator for the hydrologic cycle, therefore, but also affects nutrient flux as well. Moreover, the forest ecosystem exercises a strong regulatory function over the chemistry of liquid water passing through it. This is achieved because the structure of the ecosystem causes all precipitation to come into intimate contact with, first, the living and dead biomass and then the minerals and organic matter in the soil profile. Rapid exchange reactions in each of these compartments govern the chemical composition of liquid water leaving the ecosystem.

(8) Evapotranspiration reduces the volume of liquid water in the ecosystem. Streamwater concentrations could be increased by 1.6 times in this way. If evapotranspiration were the only operational factor, the output (concentration $\times$ amount of water) would equal the input ($1.33 \times 10^3$ Eq/ha·yr). Instead, the ecosystem is much more complex. Nutrients are released internally by weathering, accumulated by vegetation, etc. The result at Hubbard Brook is that

the ionic strength of stream water (0.20 mEq/l) is twice that of precipitation (0.10 mEq/l) and the average output is $1.65 \times 10^3$ Eq/ha·yr, whereas the average input is $1.35 \times 10^3$ Eq/ha·yr. In addition, the proportions of dissolved species change from precipitation to streamflow, further illustrating the importance of biologic and chemical reactions occurring within the ecosystem.

Hydrogen ion and sulfate dominate precipitation (pH $\sim$ 4), whereas calcium and sulfate dominate stream-water (pH $\sim$ 5) chemistry. Nitrate input in precipitation is only twice as large on an equivalent basis as ammonium but in stream-water output the proportion increases to 14 : 1.

(9) Stream-water chemistry of the undisturbed forest ecosystem is highly predictable. Concentrations of $Na^+$ and $SiO_2$ are predictably diluted as streamflow rates increase, whereas $Al^{3+}$, $NO_3^-$, $H^+$, $K^+$, and dissolved organic carbon are increased as streamflow increases. However, changes in concentrations are relatively small in relation to the great range in flow rates. It is well known that some fixed time series for sampling may seriously underestimate or overestimate a highly variable parameter (e.g., Claridge, 1970); however, most of the dissolved chemicals in stream water at the HBEF have such constant concentrations that serious errors are not produced by weekly or even monthly sampling in some cases (Johnson et al., 1969; Likens et al., 1967). Concentrations of both nitrate and potassium in stream water are highly dependent on seasonal biologic activity within the ecosystem. The biogeochemical reaction rates within these ecosystems are rapid; therefore, stream-water chemistry quickly and clearly reflects environmental conditions within the forest ecosystem.

(10) As a consequence of concentration–flow rate relations of streams, the output of most individual nutrients and the total load of dissolved substances can be closely predicted from the annual output of water alone. The few calculated values in Table 10 are therefore very reliable estimates, because hydrology has been measured and only the chemical concentration has been estimated.

(11) The undisturbed, aggrading northern hardwood ecosystem is very resistant to particulate matter erosion. This is because of its geology (stoney till) and highly

efficient mechanisms associated with living and dead biomass. Particulate matter erosion is highly dependent on flow rate, in contrast to dissolved substance removal, which is independent of flow rate. Occasional, individual storms with high discharges do most of the erosion and transport of particulate matter. In contrast to all of the other nutrients, the bulk of iron and phosphorus is lost from watershed ecosystems at Hubbard Brook in the particulate form. The stream-water output for these two elements therefore may reflect more closely the rate of stream discharge than the amount of annual streamflow. Samples of stream water obtained according to a standard time series therefore may be less accurate in estimating stream-water concentrations of iron and phosphorus than for substances occurring primarily in the dissolved form. Because discharge affects the concentration of dissolved substances and particulate matter differently, any management of water quality or total export of these materials requires different approaches at different levels of discharge.

(12) In particulate matter removal, bedload accounts for slightly more than 50% of the material removed. Many studies only measure suspended matter and hence tend to underestimate the total output of particulate matter.

(13) Dissolved substances losses from the ecosystem (110 kg of ionic substances, 38 kg of dissolved silica, and 17 kg of dissolved organic matter per hectare·year) are about five times greater than particulate matter losses (11 kg of organic matter plus 22 kg of inorganic particles). Total losses of dissolved substances and particulate matter (198 kg/ha·yr) in stream water are about 1.5 times greater than total inputs in bulk precipitation (134 kg/ha·yr).

(14) The lowering of the Hubbard Brook landscape, which is dominated by undisturbed forests, is primarily caused by solution losses and, to a lesser extent, by particulate matter losses. Slow mass movements, such as soil creep, may deliver materials to stream beds, where they are removed. At higher elevations and with steeper slopes, debris avalanches also may play a major role in lowering the landscape. Another factor that may be of considerable importance is disturbances that temporarily destroy the undisturbed forest and make it more vulnerable to both solution and particulate matter losses.

(15) In establishing input–output budgets for individual nutrients both direction of change, i.e., whether there is a net loss or gain, and magnitude of change are important. Quantitative data from six adjacent watersheds provided estimates of variability for annual budgets with time. For some ions, these budget parameters may be established with one or a few years of data, for example, direction: $Ca^{2+}(-)$, $Mg^{2+}(-)$, $Na^+(-)$, $Al^{3+}(-)$, $NH_4^+(+)$, $H^+(+)$, $SO_4^{2-}(-)$, $PO_4^{3-}(+)$, dissolved silica $(-)$, $HCO_3^-(-)$. For other ions, $K^+$, $NO_3^-$ and $Cl^-$, one or a few years gave unreliable results for direction and magnitude. For some budgets, therefore, longer term studies are necessary to obtain valid budgetary data.

(16) Undisturbed aggrading ecosystems at Hubbard Brook are accumulating 14.2 kg/ha·yr of nitrogen by microbial nitrogen fixation or other inputs of nitrogenous gases or aerosols according to estimates made by the small watershed technique. Field studies, using the acetylene reduction method, indicate vigorous nitrogen fixation in dead wood.

(17) A sulfur balance established using the Hubbard Brook ecosystem model and data gathered by the small watershed technique indicate that substantial amounts of gaseous or aerosol sulfur (6.1 kg/ha·yr) are entering the Hubbard Brook ecosystem, even though these watersheds are located more than 100 km from major polluting industries or urban areas.

(18) Three input–output patterns emerge when small watershed data are examined on a monthly basis: output exceeds input in every month for $Mg^{2+}$, $Ca^{2+}$, $Na^+$ and $Al^{3+}$; output is less than input for $NH_4^+$, $H^+$, and $PO_4^{3-}$; and crossover patterns, where for a time output exceeds input and then output is less than input, occur for $K^+$, $SO_4^{2-}$, $Cl^-$, and $NO_3^-$. These patterns are distributed among elements with both sedimentary and atmospheric cycles and reflect a complex interaction involving such factors as precipitation input, biologic activity, and climatic variations.

(19) Not only must net losses of ions from the ecosystem be included in a calculation of weathering rates, but those built into any net accumulation of biomass must also be included. At Hubbard Brook failure to include annual biomass accretion of nutrients results in a weathering estimate that is 50% too low.

(20) Chemical weathering of minerals within the ecosystem is driven by protons derived from hydrogen ions

in precipitation and those generated internally within the system by biogeochemical reactions involving carbon, nitrogen, and sulfur. The cationic weathering rate is about $2.0 \times 10^3$ Eq/ha·yr. We calculate that about 50% of the $H^+$ ions derived during the weathering reaction is derived from external sources, i.e., acid precipitation, and 50% is from internal biogeochemical sources. However, this estimate is based on the assumption that all $H^+$ ions, external plus internal, are consumed in weathering reactions.

(21) When compared to the percentage occurrence of elements in the weathering substrate, weathering estimates, indicate that differential weathering is occurring. Most rapid to least rapid are Ca > Na > Mg > K > Al > Si. The ecosystem is becoming relatively enriched in reverse order.

(22) Complete nutrient budgets for the Hubbard Brook ecosystem reveal certain basic features of nutrient cycling not otherwise apparent. These are: (a) Bulk precipitation and gaseous input are major sources of nutrients, particularly for sulfur, nitrogen, chloride, and phosphorus. (b) Weathering is the major source for Ca, Mg, K, and Na. (c) Biologic activity, photosynthesis, and nitrogen fixation, play a major role in the input of carbon and nitrogen and are important (impaction surfaces plus gaseous uptake) for sulfur. (d) The aggrading northern hardwood forest shows absolute gains for N, S, P, and Cl and losses for Si, Ca, Na, Al, Mg, and K. The latter losses are made up by weathering of primary minerals within the ecosystem. Weathering substrate may be decreasing with time.

(23) Calcium, an example of a sedimentary cycle, tends to be held within the system, with net losses being but a small proportion of the calcium on the exchange sites, 2%. However, net losses are 19% as large as annual uptake by green plants. Although sulfur has a gaseous cycle, some 99% of the sulfur in the ecosystem is found in the soil. The ecosystem approach provides an estimate of net gaseous uptake and aerosol deposition of sulfur. These sources account for about 32% of the input for the system. Meteorologic inputs dominate, in general, and release by weathering is small. Net losses of sulfur in streamflow are only 4% of the annual plant uptake. The ecosystem is much less efficient in retaining sulfur than it is in retaining calcium.

(24) Many of the important biogeochemical relationships within the forested ecosystem are regulated by microorganisms. Unfortunately, we know relatively little about the transformations mediated by microbes, such as nitrogen fixation, nitrification, and denitrification or sulfur oxidation and reduction, at Hubbard Brook. These aspects of nutrient cycling as they relate to nutrient flux must be more carefully evaluated in future studies.

(25) Input–output budgets for a diverse series of relatively undisturbed, vegetated watersheds throughout the world have many patterns in common with Hubbard Brook. Although the magnitude of precipitation, geologic substrate, vegetation type, and proximity to marine or anthropogenic emissions may vary greatly for these widespread and diverse watersheds, the constant pattern that emerges from these data is remarkable. Inputs of phosphorus and nitrogen in bulk precipitation are generally greater than losses in drainage waters on an annual basis. Conversely, outputs of calcium, magnesium, sodium, and potassium are generally greater than precipitation inputs. Net losses of potassium are small compared to inputs. The magnitude and direction for sulfate budgets are variable and depend largely on geologic substrate and intensity of air pollution.

(26) Ecosystem studies are now at the stage to provide quantitative answers for important ecological questions. The small watershed technique is useful in providing an experimental framework for these studies, which must consider geologic and meteorologic aspects as well as biologic features. Such questions include: (a) How do ecosystems change with time? Do they become more or less efficient in cycling and storing nutrients? (b) What are the effects of disturbance on the biogeochemistry of ecosystems? (c) How would weathering rates change in relation to changes in rates of biologic activity? Are there conditions under which weathering does not occur at all? (d) Of critical importance is the finding that some of the ions produced by weathering do not move downhill but are instead accumulated in the growing vegetation. If or when the forest reaches a steady state of growth, what happens to the biogeochemistry of a complex ecosystem?

# Epilog

Aldo Leopold was a very wise and perceptive observer. Some might have called him a conservationist; others might have called him an ecologist or a biogeochemist. Actually he was a brilliant, yet humble student of nature in the purest sense. Much of what this little book has described in quantitative terms Aldo Leopold outlined in general terms 25 yr ago. His insight and prose have been an inspiration to us for a long time. We therefore think it appropriate for Aldo Leopold to have the last word in our book and so, from *A Sand County Almanac* (p. 104–108) (1949), the following:

**Odyssey**   X had marked time in the limestone ledge since the Paleozoic seas covered the land. Time, to an atom locked in a rock, does not pass.

The break came when a bur-oak root nosed down a crack and began prying and sucking. In the flush of a century the rock decayed, and X was pulled out and up into the world of living things. He helped build a flower, which became an acorn, which fattened a deer, which fed an Indian, all in a single year.

From his berth in the Indian's bones, X joined again in chase and flight, feast and famine, hope and fear. He felt these things as changes in the little chemical pushes and pulls that tug timelessly at every atom. When the Indian took his leave of the prairie, X moldered briefly underground, only to embark on a second trip through the bloodstream of the land.

This time it was a rootlet of bluestem that sucked him up and lodged him in a leaf that rode the green billows of the prairie June, sharing the common task of hoarding sunlight. To this leaf also fell an uncommon task: flicking shadows across a plover's eggs. The ecstatic plover, hovering overhead, poured praises on something perfect: perhaps the eggs, perhaps the shadows, or perhaps the haze of pink phlox that lay on the prairie.

When the departing plovers set wing for the Argentine, all the bluestems waved farewell with tall new tassels. When the first geese came out of the north and all the bluestems glowed wine-red, a forehanded deermouse cut the leaf in which X lay, and buried it in an underground nest, as if to hide a bit of Indian summer from the thieving frosts. But a fox detained the mouse, molds and fungi took the nest apart, and X lay in the soil again, foot-loose and fancy-free.

Next he entered a tuft of side-oats grama, a buffalo, a buffalo chip, and again the soil. Next a spiderwort, a rabbit, and an owl. Thence a tuft of sporobolus.

All routines come to an end. This one ended with a prairie fire, which reduced the prairie plants to smoke, gas, and ashes. Phosphorus and potash atoms stayed in the ash, but the nitrogen atoms were gone with the wind. A spectator might, at this point, have predicted an early end of the biotic drama, for with fires exhausting the nitrogen, the soil might well have lost its plants and blown away.

But the prairie had two strings to its bow. Fires thinned its grasses, but they thickened its stand of leguminous herbs: prairie clover, bush clover, wild bean, vetch, lead-plant, trefoil, and Baptisia, each carrying its own bacteria housed in nodules on its rootlets. Each nodule pumped nitrogen out of the air into the plant, and then ultimately into the soil. Thus the prairie savings bank took in more nitrogen from its legumes than it paid out to its fires. That the prairie is rich is known to the humblest deermouse; why the prairie is rich is a question seldom asked in all the still lapse of ages.

Between each of his excursions through the biota, X lay in the soil and was carried by the rains, inch by inch, downhill. Living plants retarded the wash by impounding atoms; dead plants by locking them to their decayed tissues. Animals ate the plants and carried them briefly uphill or downhill, depending on whether they died or defecated higher or lower than they fed. No animal was aware that the altitude of his death was more important than his manner of dying. Thus a fox caught a gopher in a meadow, carrying X uphill to his bed on the brow of a ledge, where an eagle laid him low. The dying fox sensed the end of his chapter in foxdom, but not the new beginning in the odyssey of an atom.

An Indian eventually inherited the eagle's plumes, and with them propitiated the Fates, whom he assumed had a special interest in Indians. It did not occur to him that they might be busy casting dice against gravity; that mice and men, soils and songs, might be merely ways to retard the march of atoms to the sea.

One year, while X lay in a cottonwood by the river, he was eaten by a beaver, an animal that always feeds higher than he dies. The beaver starved when his pond dried up during a bitter frost. X rode the carcass down the spring freshet, losing more altitude each hour than heretofore in a century. He ended up in the silt of a backwater bayou, where he fed a crayfish, a coon, and then an Indian, who laid him down to his last sleep in a mound on the riverbank. One spring an oxbow caved the bank, and after one short week of freshet X lay again in his ancient prison, the sea.

An atom at large in the biota is too free to know freedom; an atom back in the sea has forgotten it. For every atom lost to the sea, the prairie pulls another out of the decaying rocks. The only certain truth is that its creatures must suck hard, live fast, and die often, lest its losses exceed its gains.

*       *       *

It is the nature of roots to nose into cracks. When Y was thus released from the parent ledge, a new animal had arrived and begun redding up the prairie to fit his own notions of law and order. An oxteam turned the prairie sod, and Y began a succession of dizzy annual trips through a new grass called wheat.

The old prairie lived by the diversity of its plants and animals, all of which were useful because the sum total of their co-operations and competitions achieved continuity. But the wheat farmer was a builder of categories; to him only wheat and oxen were useful. He saw the useless pigeons settle in clouds upon his wheat, and shortly cleared the skies of them. He saw the chinch bugs take over the stealing job, and fumed because here was a useless thing too small to kill. He failed to see the downward wash of over-wheated loam, laid bare in spring against the pelting rains. When soil-wash and chinch bugs finally put an end to wheat farming, Y and his like had already traveled far down the watershed.

When the empire of wheat collapsed, the settler took a leaf from the old prairie book: he impounded his fertility in livestock, he augmented it with nitrogen-pumping alfalfa, and he tapped the lower layers of the loam with deep-rooted corn.

But he used his alfalfa, and every other new weapon against wash, not only to hold his old plowings, but also to exploit new ones which, in turn, needed holding.

So, despite alfalfa, the black loam grew gradually thinner. Erosion engineers built dams and terraces to hold it. Army engineers built levees and wing-dams to flush it from the rivers. The rivers would not flush, but raised their beds instead, thus choking navigation. So the engineers built pools like gigantic beaver ponds, and Y landed in one of these, his trip from rock to river completed in one short century.

On first reaching the pool, Y made several trips through water plants, fish, and waterfowl. But engineers build sewers as well as dams, and down them comes the loot of all the far hills and the sea. The atoms that once grew pasque-flowers to greet the returning plovers now lie inert, confused, imprisoned in oily sludge.

Roots still nose among the rocks. Rains still pelt the fields. Deermice still hide their souvenirs of Indian summer. Old men who helped destroy the pigeons still recount the glory of the fluttering hosts. Black and white buffalo pass in and out of red barns, offering free rides to intinerant atoms.

# References

Anderson, D. H., and H. E. Hawkes. 1958. Relative mobility of the common elements in weathering of some schist and granite areas. *Geochim. Cosmochim. Acta,* **14**(3):204–210.

Aref́yeva, Z. N., and B. D. Kolesnikof. 1964. Chemistry and biochemistry dynamics of ammonia and nitrate nitrogen in forest soils of the Transurals at high and low temperatures. *Sov. Soil Sci.,* **3**:246–260.

Art, H. W., F. H. Bormann, G. K. Voigt, and G. M. Woodwell. 1974. Barrier Island forest ecosystem: the role of meteorological nutrient inputs. *Science,* **184**:60–62.

Barrett, E., and G. Brodin. 1955. The acidity of Scandinavian precipitation. *Tellus,* **7**:251–257.

Beamish, R. J. 1976. Effects of precipitation on Canadian lakes. pp. 479–498 In: *Proc. of the First International Symposium on Acid Precipitation and the Forest Ecosystem.* L. S. Dochinger and T. A. Seliga (eds). U.S.D.A. For. Serv. Gen. Tech. Rep. NE-23.

Billings, M. P. 1956. *The Geology of New Hampshire.* Part II. *Bedrock Geology.* New Hampshire State Planning and Development Commission, Concord, NH. 200 pp.

Bolin, B. (ed.). 1971. *Report of the Swedish Preparatory Committee for the United Nations Conference on Human Environment.* Norstedt and Söner, Stockholm, Sweden.

Bormann, F. H. 1974. Acid rain and the environmental future. *Environ. Cons.,* p. 270.

Bormann, F. H., and G. E. Likens. 1967. Nutrient cycling. *Science,* **155**:424–429.

Bormann, F. H., G. E. Likens, and J. S. Eaton. 1969. Biotic regulation of particulate and solution losses from a forested ecosystem. *BioScience,* **19**(7):600–610.

Bormann, F. H., T. G. Siccama, G. E. Likens, and R. H. Whittaker. 1970. The Hubbard Brook Ecosystem Study: composition and dynamics of the tree stratum. *Ecol. Monogr.,* **40**(4):373–388.

Bormann, F. H., G. E. Likens, T. G. Siccama, R. S. Pierce, and J. S. Eaton. 1974. The effect of deforestation on ecosystem export and the steady-state condition at Hubbard Brook. *Ecol. Monogr.,* **44**(3):255–277.

Bormann, F. H., G. E. Likens, and J. Melillo. 1977. Nitrogen budget for an aggrading northern hardwood forest ecosystem. Submitted to *Science.*

Bradley, E., and R. V. Cushman. 1956. Memorandum report on geologic and ground-water conditions in the Hubbard Brook watershed, New Hampshire. On file at Northeastern Forest Experiment Station. Durham, NH. 15 pp.

Brady, N. C. 1974. *The Nature and Properties of Soils.* Macmillan, New York. 639 pp.

Braun, E. L. 1950. *Deciduous Forests of Eastern North America.* The Blakiston Co., Philadelphia, PA. 594 pp.

Burton, T. M., and G. E. Likens. 1975. Energy flow and nutrient cycling in salamander populations in the Hubbard Brook Experimental Forest, New Hampshire. *Ecology,* **56**(5):1068–1080.

Carlisle, A., A. H. F. Brown, and E. J. White. 1967. The nutrient content of tree stem flow and ground flora litter and leachates in a sessile oak *(Quercus petraea)* woodland. *J. Ecol.,* **55**:615–627.

Carroll, D. 1970. *Rock Weathering.* Plenum Press, New York, NY. 203 pp.

Claridge, G. G. C. 1970. Studies in element balances in a small catchment at Taita, New Zealand. *Proceedings of the International Association of Scientific Hydrology, UNESCO Symp.,* on Results of Research on Representative and Experimental Basins. Pp. 523–540. Wellington, New Zealand, December 1970.

Cogbill, C. V. 1976. The effect of acid precipitation on tree growth in eastern North America. pp. 1027–1032. In: *Proceedings of the First International Symposium on Acid Precipitation and the Forest Ecosystem.* L. S. Dochinger, and T. A. Seliga (eds.) U.S.D.A For. Serv. Gen. Tech. Rep. NE-23.

Cogbill, C. V., and G. E. Likens. 1974. Acid precipitation in the northeastern United States. *Water Resources Res.,* **10**:1133–1137.

Cole, D. W., S. P. Gessel, and S. F. Dice. 1967. Distribution and cycling of nitrogen, phosphorus, potassium, and calcium in a second-growth douglas-fir ecosystem. *Symposium on Primary Productivity and Mineral Cycling in Natural Ecosystems.* New York City, December 27, 1967, University of Maine Press, Orono, ME. 245 pp.

Conrad, V. 1941. The variability of precipitation. *Mon. Weath. Rev. Wash.*, **69**:5.

Covington, W. W. 1975. Successional dynamics of organic matter and nutrient content of forest floor and leaf fall in northern hardwoods. *Bull. Ecol. Soc. Amer.*, **56**(2):12.

Covington, W. W. 1976. Secondary succession in northern hardwoods: forest floor organic matter and nutrients and leaf fall. Ph.D. Thesis, Yale University, New Haven, CT. 117 pp.

Cowling, D. W., L. H. P. Jones, and D. R. Lockyer, 1973. Increased yield through correction of sulphur deficiency in ryegrass exposed to sulphur dioxide. *Nature,* **243**(5407):479–480.

Crisp, D. T. 1966. Input and output of minerals for an area of pennine moorland: the importance of precipitation, drainage, peat erosion and animals. *J. Appl. Ecol.*, **3**:327–348.

Croat, T. B. 1972. The role of overpopulation and agricultural methods in the destruction of tropical ecosystems. *BioScience*, **22**(8):465–467.

Cryer, R. 1976. The significance and variation of atmospheric nutrient inputs in a small catchment system. *J. Hydrol.*, **29**:121–137.

Dominski, A. S. 1971. Nitrogen transformations in a northern-hardwood podzol on cutover and forested sites. Ph.D. Thesis, Yale University, New Haven, CT. 157 pp.

Douglass, J. E., and W. T. Swank. 1972. Streamflow modification through management of Eastern forests. Southeast Forest Experiment Station, U.S.D.A. Forest Service Research Paper SE-94, Washington, DC. 15 pp.

Duvigneaud, P., and S. Denaeyer-DeSmet. 1964. Le cycle des éléments biogènes dans l'écosystème forêt (Forêts tempèrées caducifoliées). *Lejeunia,* **28**:1–148.

Eaton, J. S., G. E. Likens, and F. H. Bormann. 1973. Throughfall and stemflow chemistry in a northern hardwood forest. *J. Ecol.*, **61**:495–508.

Eaton, J. S., G. E. Likens, and F. H. Bormann. 1976. The biogeochemistry of sulfur in a northeastern forest ecosystem. p. 43. In: *The Tenth Middle Atlantic Regional Meeting Abstracts,* American Chemical Society, February 23–26, Philadelphia, PA.

Eriksson, E. 1952. Composition of atmospheric precipitation. II. Sulfur, chloride, iodine compounds. Bibliography. *Tellus,* **4**:280–303.

Eriksson, E. 1974. Vattnet-Kemikaliebäraren. *Forskning och Framsteg,* **5**:41–45.

Federer, C. A. 1973. Annual cycles of soil and water temperatures at Hubbard Brook. United States Forest Service Research Note NE-167. Upper Darby, PA. 7 pp.

Fisher, D. W., A. W. Gambell, G. E. Likens, and F. H. Bormann. 1968. Atmospheric contributions to water quality of streams in the Hubbard Brook Experimental Forest, New Hampshire. *Water Resources Res.,* **4**(5):1115–1126.

Fisher, S. G. 1970. Annual energy budget of a small forest stream ecosystem: Bear Brook, West Thornton, New Hampshire. Ph.D. Thesis, Dartmouth College, Hanover, NH. 97 pp.

Fisher, S. G., and G. E. Likens, 1973. Energy flow in Bear Brook, New Hampshire: an integrative approach to stream ecosystem metabolism. *Ecol. Monogr.,* **43**(4):421–439.

Flaccus, E. 1958a. Landslides and their revegetation in the White Mountains of New Hampshire. Ph.D. dissertation, Duke University, Durham, NC. 186 pp.

Flaccus, E. 1958b. White Mountain landslides. *Appalachia,* **32**:175–191.

Fowler-Billings, K., and L. R. Page. 1942. The Geology of the Cardigan and Rumney Quadrangles, New Hampshire. New Hampshire Planning and Development Commission. Concord, NH. 31 pp.

Fredriksen, R. L. 1975. Nitrogen, phosphorus and particulate matter budgets of five coniferous forest ecosystems in the Western Cascades Range, Oregon, Ph.D. Thesis, Oregon State University, Corvallis, OR.

Galloway, J. N., G. E. Likens, and E. S. Edgerton. 1976. Acid precipitation in the northeastern United States: pH and acidity. *Science* **194**:722–724.

Gambell, A. W., and D. W. Fisher. 1966. Chemical composition of rainfall of eastern North Carolina and southeastern Virginia. United States Geological Survey Water Supply Paper 1535-K. Washington, DC. pp. K1–K41.

Gjessing, E. T., A. Henriksen, M. Johannessen, and R. F. Wright. 1976. Effects of acid precipitation on freshwater chemistry. Pp. 65–85. In: F. H. Braekke (ed.). *Impact of Acid Precipitation on Forest and Freshwater Ecosystems in Norway.* Report FR 6/76, SNSF-Project, Oslo, Norway.

Gómez-Pompa, A., C. Vázquez-Yanes, and S. Guevara. 1972. The tropical rainforest: a nonrenewable resource. *Science,* **177**:762–765.

Goldich, S. S. 1938. A study of rock weathering. *J. Geol.,* **46**:17–58.

Gorham, E. 1958. Atmospheric pollution by hydrochloric acid. *Quart. J. Roy. Meteorol. Soc.,* **84**(361):274–276.

Gosz, J. R., G. E. Likens, and F. H. Bormann. 1972. Nutrient content of litter fall on the Hubbard Brook Experimental Forest, New Hampshire. *Ecology,* **53**(5):769–784.

Gosz, J. R., G. E. Likens, and F. H. Bormann. 1973. Nutrient release from decomposing leaf and branch litter in the Hubbard Brook Forest, New Hampshire. *Ecol. Monogr.,* **43**(2):173–191.

Gosz, J. R., G. E. Likens, and F. H. Bormann. 1976. Organic

matter and nutrient dynamics of the forest and forest floor in the Hubbard Brook Forest. *Oecologia* (Berlin), **22**:305–320.

Gosz, J. R. 1977. Nutrient budgets in relation to climax, succession and productivity in New Mexico. *Ecology* (In Press).

Granat, L. 1972. On the relation between pH and the chemical composition in atmospheric precipitation. *Tellus*, **24**:550–560.

Hack, J. T. and J. C. Goodlett. 1960. Geomorphology and forest ecology of a mountain region in the Central Appalachians. United States Geological Survey Professional Paper 347. Washington, DC. 66 pp.

Hamon, W. R. 1961. Estimating potential evapotranspiration. *Amer. Soc. Civ. Eng. Proc.*, **87**(HY3):107–120.

Hart, G. E. Jr. 1966. Streamflow characteristics of small forested watersheds in the White Mountains of New Hampshire. Ph.D. Thesis, University of Michigan. Ann Arbor, MI. 141 pp.

Hart, G. E. Jr., R. E. Leonard, and R. S. Pierce. 1962. Leaf fall, humus depth, and soil frost in a northern hardwood forest. Research Note No. 131, Northeastern Forest Experiment Station, Upper Darby, PA. 3 pp.

Herman, F. A. and E. Gorham. 1957. Total mineral material, acidity, sulfur, and nitrogen in rain and snow at Kentville, Nova Scotia. *Tellus*, **9**:180–183.

Hill, A. C. 1971. Vegetation: a sink for atmospheric pollutants. *J. Air Poll. Contr. Assoc.*, **21**:341–346.

Hobbie, J. E. and G. E. Likens. 1973. The output of phosphorus dissolved organic carbon and fine particulate carbon from Hubbard Brook watersheds. *Limnol. Oceanogr.*, **18**(5):734–742.

Hoeft, R. G., D. R. Keeney, and L. M. Walsh. 1972. Nitrogen and sulfur in precipitation and sulfur dioxide in the atmosphere in Wisconsin. *J. Environ. Qual.*, **1**:203–208.

Holmes, R. T., and F. W. Sturges. 1973. Annual energy expenditure by the avifauna of a northern hardwoods ecosystem. *Oikos*, **24**:24–29.

Hornbeck, J. W., R. S. Pierce, and C. A. Federer. 1970. Streamflow changes after forest clearing in New England. *Water Resources Res.*, **6**(4):1124–1132.

Hornbeck, J. W., G. E. Likens, R. S. Pierce, and F. H. Bormann. 1975a. Strip cutting as a means of protecting site and streamflow quality when clearcutting northern hardwoods. Pp. 209–229. In: B. Bernier and C. H. Winget (eds.), Proceedings of the 4th North American Forest Soils Conference on Forest Soils and Forest Land Management, August 1973, Quebec, Canada. *Forest Soils and Forest Land Management*, University of Laval Press.

Hornbeck, J. W., R. S. Pierce, G. E. Likens, and C. W. Martin. 1975b. Moderating the impact of contemporary forest cutting on hydrologic and nutrient cycles. Pp. 423–433. In: *Interna-*

*tional Symposium on the Hydrological Characteristics of River Basins.* Tokyo, December 1975, International Association of Hydrological Sciences Publ. 117.

Hornbeck, J. W., G. E. Likens, and J. S. Eaton. 1976. Seasonal variation in acidity of precipitation and the implications for forest-stream ecosystems. pp. 597–609. In: *Proceedings of the First International Symposium on Acid Precipitation and the Forest Ecosystem.* L. S. Dochinger and T. A. Seliga (eds.) U.S.D.A. For. Serv. Gen. Tech. Rep. NE-23.

Hunt, C. B. 1967. Physiography of the United States. Freeman and Co., San Francisco, CA.

Hutchinson, G. E. 1957. *A Treatise on Limnology.* Vol. 1. J. Wiley and Sons, Inc. New York, N.Y. 1015 pp.

Johannessen, M., T. Dale, E. T. Gjessing, A. Henriksen, and R. F. Wright. 1976. *Proceedings of the International Symposium on Isotopes and Impurities in Snow and Ice.* International Association of Hydrological Science, Grenoble, France, August 28–30, 1975. International Association of Hydrological Science Publ. 118.

Johnson, N. M. 1971. Mineral equilibria in ecosystem geochemistry. *Ecology,* **52**:529–531.

Johnson, N. M., G. E. Likens, F. H. Bormann, and R. S. Pierce. 1968. Rate of chemical weathering of silicate minerals in New Hampshire. *Geochim. Cosmochim. Acta,* **32**:531–545.

Johnson, N. M., G. E. Likens, F. H. Bormann, D. W. Fisher, and R. S. Pierce. 1969. A working model for the variation in streamwater chemistry at the Hubbard Brook Experimental Forest, New Hampshire. *Water Resources Res.,* **5**(6):1353–1363.

Johnson, N. M., R. C. Reynolds, and G. E. Likens. 1972. Atmospheric sulfur: its effect on the chemical weathering of New England. *Science,* **177**:514–516.

Johnson, P. L., and W. T. Swank. 1973. Studies of cation budgets in the southern Appalachians on four experimental watersheds with contrasting vegetation. *Ecology,* **54**(1):70–80.

Jordan, M. J., and G. E. Likens. 1975. An organic carbon budget for an oligotrophic lake in New Hampshire, U.S.A. *Verh. Intl. Verein. Limnol.,* **19**(2):994–1003.

Jordan, C. F., and J. R. Kline. 1972. Relative stability of mineral cycles in forest ecosystems. *Amer. Nat.,* **106**(948):237–253.

Juang, F. H. T., and N. M. Johnson. 1967. Cycling of chlorine through a forested catchment in New England. *J. Geophys. Res.,* **72**:5641–5647.

Junge, C. E. 1958. The distribution of ammonia and nitrate in rain water over the United States. *Trans. Amer. Geophys. Union,* **39**:241–248.

Junge, C. E. 1963. *Air Chemistry and Radioactivity.* Academic Press, New York, NY. 382 pp.

Junge, C. E., and R. T. Werby. 1958. The concentration of chloride, sodium, potassium, calcium and sulfate in rain water over the United States. *J. Meteorol.,* **15**:417–425.

Küchler, A. W. 1964. Potential natural vegetation of the conterminous United States. American Geographical Society Spec. Publ. No. 36. New York, NY. 116 pp.

Leonard, R. E. 1961. Interception of Precipitation by Northern Hardwoods. Station Paper No. 159, Northeastern Forest Experiment Station, United States Forest Service, Upper Darby, PA. 16 pp.

Leopold, A. 1949. *A Sand County Almanac.* Oxford University Press, Oxford, England. 226 pp.

Likens, G. E. 1972. *The Chemistry of Precipitation in the Central Finger Lakes Region.* Tech. Publ. No. 50, Cornell University Water Resources and Marine Sciences Center, Ithaca, New York. 47 pp. and 14 Figs.

Likens, G. E. 1973. *A Checklist of Organisms for the Hubbard Brook Ecosystems.* Section of Ecology and Systematics, Cornell University, Ithaca, NY. Mimeo. 54 pp.

Likens, G. E., and F. H. Bormann. 1970. *Chemical Analyses of Plant Tissues from the Hubbard Brook Ecosystem in New Hampshire.* Bull. 79, Yale University School of Forestry, New Haven, CT. 25 pp.

Likens, G. E., and F. H. Bormann. 1972. Nutrient cycling in ecosystems. Pp. 25–67. In: J. Wiens (ed.), *Ecosystems Structure and Function.* Oregon State University Press, Corvallis, OR.

Likens, G. E., and F. H. Bormann. 1974a. Linkages between terrestrial and aquatic ecosystems. *BioScience,* **24**(8):447–456.

Likens. G. E., and F. H. Bormann. 1974b. Acid rain: a serious regional environmental problem. *Science,* **184**(4143):1176–1179.

Likens, G. E., and F. H. Bormann. 1975. An experimental approach to nutrient-hydrologic interactions in New England landscapes. Pp. 7–29. In: A. D. Hasler (ed.), *Proceedings of the INTECOL Symposium on Coupling of Land and Water Systems, 1971.* Leningrad. Springer-Verlag, New York, NY.

Likens, G. E., and M. B. Davis. 1975. Post-glacial history of Mirror Lake and its watershed in New Hampshire, U.S.A.: an initial report. *Verh. Intl. Verein. Limnol.,* **19**:982–993.

Likens, G. E., F. H. Bormann, N. M. Johnson, and R. S. Pierce. 1967. The calcium, magnesium, potassium and sodium budgets for a small forested ecosystem. *Ecology,* **48**:772–785.

Likens, G. E., F. H. Bormann, N. M. Johnson, D. W. Fisher, and R. S. Pierce. 1970. The effect of forest cutting and herbicide treatment on nutrient budgets in the Hubbard Brook watershed-ecosystem. *Ecol. Monogr.,* **40**(1):23–47.

Likens, G. E., F. H. Bormann, R. S. Pierce, and D. W. Fisher.

1971. Nutrient–hydrologic cycle interaction in small forested watershed-ecosystems. Pp. 553–563. In: P. Duvigneaud (ed.), *Productivity of Forest Ecosystems,* Proceedings of the Brussels Symposium, 1969. UNESCO, Paris, France.

Likens, G. E., F. H. Bormann, and N. M. Johnson. 1972. Acid rain. *Environment,* **14**(2):33–40.

Likens, G. E., N. M. Johnson, J. N. Galloway, and F. H. Bormann. 1976a. Acid precipitation: strong and weak acids. *Science,* **194**(4265):643–645.

Likens, G. E., F. H. Bormann, J. S. Eaton, R. S. Pierce, and N. M. Johnson. 1976b. Hydrogen ion input to the Hubbard Brook Experimental Forest, New Hampshire during the last decade. pp. 397–407. In: *Proceedings of the First International Symposium on Acid Precipitation and the Forest Ecosystem,* L. S. Dochinger and T. A. Seliga (eds.) U.S.D.A. For. Serv. Gen. Tech. Rep. NE-23.

Lunt, H. A. 1932. Profile characteristics of New England forest soils. *Conn. Agri. Exp. Sta. Bull.,* **342**:743–836.

MacIntyre, F. 1974. The top millimeter of the ocean. *Sci. Amer.,* **230**:67–77.

Marks, P. L. 1974. The role of pin cherry (*Prunus pensylvanica* L.) in the maintenance of stability in northern hardwood ecosystems. *Ecol. Monogr.,* **44**(1):73–88.

Marshall, C. E. 1964. *The physical chemistry and mineralogy of soils.* Vol. 1. *Soil Materials.* Wiley, New York, NY. 388 pp.

McGarity, J. W., and J. A. Rajaratnam. 1973. Apparatus for the measurement of losses of nitrogen as gas from the field and simulated field environments. *Soil. Biol. Biochem.,* **5**:121–131.

Melillo, J. M. 1977. Nitrogen dynamics in an aggrading northern hardwood forest ecosystem. Ph.D. Thesis, Yale University, New Haven, CT.

Miller, R. B. 1963. Plant nutrients in hard beech—the immobilization of nutrients. *New Zealand J. Sci.,* **6**:365–377.

Moke, C. B. 1946. The geology of the Plymouth quadrangle, New Hampshire. New Hampshire Planning and Development Commission. Concord, NH, 21 pp.

Muller, R. N., and F. H. Bormann. 1976. The role of *Erythronium americanum* Ker. in energy flow and nutrient dynamics of a northern hardwood forest ecosystem. *Science,* **193**:1126–1128.

National Academy of Sciences. 1974. *United States Participation in the International Biological Program.* Rept. No. 6. Washington, DC. 166 pp.

Northeastern Forest Experiment Station, U.S. Forest Service. 1972. Hubbard Brook Experimental Forest. Northeastern Forest Experiment Station, Upper Darby, PA. 13 pp.

Odum, E. P. 1959. *Fundamentals of Ecology.* 2nd Ed. W. B. Saunders Co., Philadelphia, PA. 546 pp.

Oosting, H. J. 1956. *The Study of Plant Communities*. W. H. Freeman Co., San Francisco, CA. 440 pp.

Pearson, F. J., and D. W. Fisher. 1971. Chemical Composition of Atmospheric Precipitation in the Northeastern United States. United States Geological Survey Water Supply Paper 1535-P. Washington, DC. pp. P1–P23.

Penman, H. L. 1956. Estimating evapotranspiration. *Trans. Amer. Geophys. Union,* **37**:43–50.

Pierce, R. S. 1967. Evidence of overland flow on forest watersheds. Pp. 247–252. In: W. E. Sopper and H. W. Lull (eds.), *International Symposium on Forest Hydrology*. Pergamon Press, New York, NY.

Pierce, R. S., J. W. Hornbeck, G. E. Likens, and F. H. Bormann. 1970. Effects of elimination of vegetation on stream water quantity and quality. Pp. 311–328. In: Proc. International Association Scientific Hydrology. UNESCO. *Results of Research on Representative and Experimental Basins,* Wellington, New Zealand, December 1970.

Pierce, R. S., C. W. Martin, C. C. Reeves, G. E. Likens, and F. H. Bormann. 1972. Nutrient loss from clearcuttings in New Hampshire. *Symposium on Watersheds in Trans.,* Ft. Collins, Colorado, Pp. 285–295. American Water Resources Assoc. and Colorado State Univ.

Powell, S. T. 1964. Quality of water. Pp. 19-1 to 19-37. In: V. T. Chow (ed.), *Handbook of Applied Hydrology*. McGraw-Hill Book Co., New York, NY.

Polynov, B. B. 1937. *The cycle of weathering*. (A. Muir, Transl.) Thomas Murby. London, England.

Reuss, J. O. 1976. Chemical and biological relationships relevant to the ecological effects of acid rainfall. Pp. 791–813. In: *Proceedings of the First International Symposium on Acid Precipitation and the Forest Ecosystem*. L. S. Dochinger and T. A. Seliga (eds.) U.S.D.A. For. Serv. Gen. Tech. Rep. NE-23.

Reynolds, R. C., and N. M. Johnson. 1972. Chemical weathering in the temperate glacial environment of the northern Cascade Mountains. *Geochim. Cosmochim. Acta,* **36**:537–554.

Roskoski, J. P. 1975. Differential nitrogen fixation in wood litter. *Bull. Ecol. Soc. Amer.,* **56**(2):12.

Schindler, D. W., R. W. Newbury, K. G. Beaty, and P. Campbell. 1976. Natural water and chemical budgets for a small Precambrian lake basin in central Canada. *J. Fish. Res. Bd. Canada* **33**:2526–2543.

Schindler, D. W., and J. E. Nighswander. 1970. Nutrient supply and primary production in Clear Lake, Eastern Ontario. *J. Fish. Res. Bd. Can.,* **27**(11):2009–2036.

Schofield, C. L. 1976. Lake acidification in the Adirondack Mountains of New York: causes and consequences. p. 477. In: *Proceedings of the First International Symposium on Acid Precipitation and the Forest Ecosystem*. L. S. Dochinger and T. A. Seliga (eds.) U.S.D.A. For. Serv. Gen. Tech. Rep. NE-23.

Scrivener, J. C. 1975. Water, water chemistry and hydrochemical balance of dissolved ions in Carnation Creek watershed, Vancouver Island, July 1971–May 1974. Technical Report 564, Fisheries Marine Services Research and Development. Vancouver, B.C. 141 pp.

Sellers, W. D. 1965. *Physical Climatology*. The University of Chicago Press, Chicago, IL. 272 pp.

Siccama, T. G., F. H. Bormann, and G. E. Likens. 1970. The Hubbard Brook Ecosystem Study: productivity, nutrients and phytosociology of the herbaceous layer. *Ecol. Monogr.* **40**(4):389–402.

Siegler, H. R. (ed.). 1968. *The White-tailed Deer of New Hampshire*. New Hampshire Fish and Game Dept., Concord, NH. 256 pp.

Smith, W. H. 1976. Character and significance of forest tree root exudates. *Ecology,* **57**:324–331.

Sopper, W. E., and H. W. Lull, 1965. The representativeness of small forested experimental watersheds in northeastern United States. *Intl. Assoc. Sci. Hydrol.,* **66**(2):441–456.

Sopper, W. E., and H. W. Lull. 1970. *Streamflow Characteristics of the Northeastern United States*. Bull. 766. The Pennsylvania State University, Univ. Park, PA. 129 pp.

Stumm, W., and J. J. Morgan. 1970. *Aquatic Chemistry*. Wiley-Interscience, New York, NY. 583 pp.

Sturges, F. W., R. T. Holmes, and G. E. Likens. 1974. The role of birds in nutrient cycling in a northern hardwoods ecosystem. *Ecology,* **55**(1):149–155.

Tamm, C. O., and T. Troedsson. 1955. An example of the amounts of plant nutrients supplied to the ground in road dust. *Oikos,* **6**:61–70.

Taylor, A. W., W. M. Edwards, and E. C. Simpson. 1971. Nutrients in streams draining woodland and farm land near Coshocton, Ohio. *Water Resources Res.,* **7**(1):81–89.

Thiessen, A. H. 1911. Precipitation for large areas. *Monthly Weather Rev.,* **39**:1082–1089.

Thornthwaite, C. W. 1948. An approach toward a rational classification of climate. *Geogr. Rev.,* **38**:55–94.

Trewartha, G. T. 1954. *Introduction to climate*. McGraw-Hill Book Co., New York, NY. 402 pp.

Ungemach, H. 1967. Pp. 221–226. In: H. Lent (ed.), *Atas do Simpósio sâbre a Biota Amazônica*. Vol. 3. Rio de Janeiro, Brazil.

Ungemach, H. 1970. Chemical rain water studies in the Amazon region. Pp. 354–358. In: J. M. Idrobo (ed.), *Simpósio Y Ford de Biologia Tropical Amazônica*. Vol. II. Editorial PAX, Bogotá, Colombia.

Viro, P. J. 1953. Loss of nutrients and the natural nutrient balance of the soil in Finland. *Comm. Inst. Forest Fenn.,* **42**:1–50.

White, E. J., and F. Turner. 1970. A method of estimating income of nutrients in catch of airborne particles by a woodland canopy. *J. Appl. Ecol.,* **7**:441–461.

Whittaker, R. H., and G. E. Likens. 1973. Carbon in the biota. Pp. 281–302. In: G. M. Woodwell and E. V. Pecan (eds.), *Carbon and the Biosphere.* CONF - 720510, United States Atomic Energy Commission, Springfield, VA.

Whittaker, R. H., and G. E. Likens. 1975. The biosphere and man. Pp. 305–328. In: H. Lieth and R. H. Whittaker (eds.), *Primary Productivity of the Biosphere.* Springer-Verlag, NY.

Whittaker, R. H., F. H. Bormann, G. E. Likens, and T. G. Siccama. 1974. The Hubbard Brook Ecosystem Study: forest biomass and production. *Ecol. Monogr.,* **44**(2):233–254.

Wood, T., and F. H. Bormann. 1974. The effects of an artificial acid mist upon the growth of *Betula alleghaniensis* Britt. *Environ. Pol.,* **7**:259–268.

Wood, T., and F. H. Bormann. 1975. Increases in foliar leaching caused by acidification of a artificial mist. *Ambio,* **4**(4):169–171.

Woodwell, G. M., and R. H. Whittaker. 1967. Primary production and the cation budget of the Brookhaven Forest. Pp. 151–166. In: H. E. Young (ed.), *Symposium on Primary Productivity and Mineral Cycling in Natural Ecosystems.* University of Maine Press, Orono, ME.

Wright, R. F. 1974. *Forest Fire: Impact on the Hydrology, Chemistry and Sediments of Small Lakes in Northeastern Minnesota.* Interim Rept. No. 10, Limnology Research Center, University of Minnesota, Minneapolis, MN. 129 pp.

Wright, R. F., T. Dale, E. T. Gjessing, G. R. Hendrey, A. Henriksen, M. Johannessen, and I. P. Muniz. 1976. Impact of acid precipitation on fresh-water ecosystems in Norway. Pp. 459–476. In: *Proceedings of the First International Symposium on Acid Precipitation and the Forest Ecosystem.* L. S. Dochinger and T. A. Seliga (eds.) U.S.D.A. For. Serv. Gen. Tech. Rep. NE-23.

# Subject Index

References to the principal treatment of a subject (section headings) are set in **boldface.**

Sulfur [*cont.*]
  net mineralization   97, 101
  output in stream water   62, 71, 97, 100–102, 111
  plant uptake   97, 100–102
  role in weathering   89–90
  weathering release   72, 97, 99–102
Sweden *(see Nutrient Budget)*
Swift River, NH   28

**T**aughannock Creek, NY, US *(see Nutrient Budget)*
Temperature
  mean air   7
  stream   7
Tesuque Watersheds, NM, US *(see Nutrient Budget)*
Thiessen Method   15
Till, Glacial   7, 9, 58
Topography   7
  pit and mound   23
  role of   58, 62–64
Throughfall   43, **45** ff, 72, 96–97

United States *(see Nutrient Budget and Watersheds)*

**V**ectors   3–5
  biologic   4
  geologic   65
  meteorologic   44–46, 65
Vegetation Type   9–11, 105–110 *(also see Nutrient Budget)*
Velen, Sweden *(see Nutrient Budget)*

**W**ales *(see Nutrient Budget)*
Walker Branch, TE, US *(see Nutrient Budget)*
Watershed-Ecosystem Approach   4–5, 114–115
Watersheds
  experimental   27–29
  Maine   27–28
  Massachusetts   27–28
  New Hampshire   27–28
  Vermont   27–28
Water Year   **14**–15
Weathering   33, 48, 65, 72, **87** ff, 95–102, 104, 119–121
  differential rates of   92–93, 120
  in New England   91–92
  in North America   91
  rates   90–92
Weir   6, 20–21, 49, 57–58
Western Cascades Range, OR, US *(see Nutrient Budget)*
White Mountains, New Hampshire   2, 5
Wind, Prevailing   7

# Ecological Studies
## Analysis and Synthesis

Edited by W. D. Billings, F. Golley, O. L. Lange, and J. S. Olson

Springer-Verlag
New York   Heidelberg   Berlin